VOLUME ONE HUNDRED AND THIRTY THREE

PROGRESS IN MOLECULAR BIOLOGY AND TRANSLATIONAL SCIENCE

RGS Protein Physiology and Pathophysiology

VOLUME ONE HUNDRED AND THIRTY THREE

PROGRESS IN MOLECULAR BIOLOGY AND TRANSLATIONAL SCIENCE

RGS Protein Physiology and Pathophysiology

Edited by

RORY A. FISHER
Department of Pharmacology, University of Iowa, Iowa City, Iowa, USA

ELSEVIER

AMSTERDAM • BOSTON • HEIDELBERG • LONDON
NEW YORK • OXFORD • PARIS • SAN DIEGO
SAN FRANCISCO • SINGAPORE • SYDNEY • TOKYO
Academic Press is an imprint of Elsevier

AP

Academic Press is an imprint of Elsevier
225 Wyman Street, Waltham, MA 02451, USA
525 B Street, Suite 1800, San Diego, CA 92101-4495, USA
125 London Wall, London, EC2Y 5AS, UK
The Boulevard, Langford Lane, Kidlington, Oxford OX5 1GB, UK

First edition 2015

Notices
Knowledge and best practice in this field are constantly changing. As new research and experience broaden our understanding, changes in research methods, professional practices, or medical treatment may become necessary.

Practitioners and researchers must always rely on their own experience and knowledge in evaluating and using any information, methods, compounds, or experiments described herein. In using such information or methods they should be mindful of their own safety and the safety of others, including parties for whom they have a professional responsibility.

To the fullest extent of the law, neither the Publisher nor the authors, contributors, or editors, assume any liability for any injury and/or damage to persons or property as a matter of products liability, negligence or otherwise, or from any use or operation of any methods, products, instructions, or ideas contained in the material herein.

ISBN: 978-0-12-802938-1
ISSN: 1877-1173

CONTENTS

CONTRIBUTORS

Kendall J. Blumer
Department of Cell Biology and Physiology, Washington University School of Medicine, St. Louis, Missouri, USA

Ching-Kang Jason Chen
Department of Ophthalmology; Department of Biochemistry and Molecular Biology, and Department of Neuroscience, Baylor College of Medicine, Houston, Texas, USA

Wei Chen
Department of Pathology, School of Medicine, University of Alabama at Birmingham, Birmingham, Alabama, USA

Serena M. Dudek
Neurobiology Laboratory, National Institute of Environmental Health Sciences, National Institutes of Health, Research Triangle Park, North Carolina, USA

Paul R. Evans
Department of Pharmacology, Emory University School of Medicine, Rollins Research Center, Atlanta, Georgia, USA

Rory A. Fisher
Department of Pharmacology, University of Iowa Carver College of Medicine, Iowa City, Iowa, USA

Ruth Ganss
Harry Perkins Institute of Medical Research, Centre for Medical Research, The University of Western Australia, Perth, Western Australia, Australia

John R. Hepler
Department of Pharmacology, Emory University School of Medicine, Rollins Research Center, Atlanta, Georgia, USA

Joel Jules
Department of Pathology, School of Medicine, University of Alabama at Birmingham, Birmingham, Alabama, USA

Jae-Kyung Lee
Department of Physiology, Emory University School of Medicine, Atlanta, Georgia, USA

Yi-Ping Li
Department of Pathology, School of Medicine, University of Alabama at Birmingham, Birmingham, Alabama, USA

Biswanath Maity
Department of Pharmacology, University of Iowa Carver College of Medicine, Iowa City, Iowa, USA

Richard R. Neubig
Department of Pharmacology & Toxicology, Michigan State University, East Lansing, Michigan, USA

Patrick Osei-Owusu
Department of Pharmacology and Physiology, Drexel University College of Medicine, Philadelphia, Pennsylvania, USA

Adele Stewart*
Department of Pharmacology, University of Iowa Carver College of Medicine, Iowa City, Iowa, USA

Malú G. Tansey
Department of Physiology, Emory University School of Medicine, Atlanta, Georgia, USA

Shuying Yang
Department of Oral Biology, School of Dental Medicine, and Developmental Genomics Group, New York State Center of Excellence in Bioinformatics and Life Sciences, University at Buffalo, The State University of New York, Buffalo, New York, USA

*Present address: Department of Pharmacology, Vanderbilt University, Nashville, Tennessee, USA.

PREFACE

RGS proteins and their first identified "physiological" role were discovered by genetic studies in yeast more than 30 years ago. In that work, loss-of-function mutations in the yeast *SST2* gene were found to promote "supersensitivity" to the pheromone α-factor, demonstrating that the novel protein encoded by *SST2* (Ss2tp) functioned to promote recovery from pheromone-induced growth arrest. Given that pheromone signaling in yeast is mediated through G protein-coupled receptors (GPCRs), these findings raised the intriguing possibility that RGS proteins, if present in humans, might play significant roles in physiology and disease. Indeed, GPCRs regulate virtually every known physiological process and are the targets of 40–50% of currently marketed pharmaceuticals. The ensuing discovery of the existence of a family of RGS proteins in higher organisms including humans incited a firestorm of interest in RGS proteins that yielded enormous advances to provide our current understanding of RGS protein function.

It is now clear that RGS proteins are multifunctional GTPase-accelerating proteins (GAPs) that serve to promote inactivation of specific Gα subunits rather than GPCRs. Because of this activity, RGS proteins determine the magnitude and duration of cellular responses initiated by many GPCRs. RGS proteins are defined by the presence of a semi-conserved 130-amino acid RGS domain whose structural features and mechanism of accelerated GTP hydrolysis by G proteins have been defined. Twenty canonical mammalian RGS proteins, divided into four subfamilies, act as functional GAPs while almost 20 additional proteins contain non-functional RGS-like domains that often mediate interactions with GPCRs or Gα subunits. Certain RGS proteins have been shown to interact with GPCRs, to act as effector antagonists and to possess G protein-independent functions. While RGS protein biochemistry and signaling has been well elucidated *in vitro*, the physiological functions of each RGS family member remain largely unexplored.

This volume of *Progress in Molecular Biology and Translational Science* summarizes recent advances employing genetically modified model organisms that provide the first insights into RGS protein functions *in vivo*. In addition, this work has provided intriguing evidence that the contribution of RGS proteins to biological outcomes *in vivo* can be as important as those initiated by activation of GPCRs. Historically, a lack of specific antibodies with corresponding genetic knockout controls made detection of endogenous RGS proteins difficult *in vivo*, making it challenging to uncover the physiological significance of RGS proteins. Moreover, the potential for functional redundancy of RGS proteins, a possibility suggested by the existence of multiple RGS transcripts that act upon the same Gα subunits in tissues, represented another challenge to investigating RGS protein function *in vivo*. Combinatorial knockout of multiple RGS proteins to investigate the net importance of RGS protein function in a particular disease or physiological process until recently has been a technical and financial nightmare. This volume devotes a chapter describing one approach to overcome these challenges by creation of mice expressing knock-in alleles of RGS-insensitive Gα mutants. In addition, this volume provides multiple examples of how individual deletion of RGS proteins, despite the potential for RGS protein redundancy, revealed striking roles for RGS proteins *in vivo* and identified RGS proteins as novel therapeutic targets for various diseases. Particularly interesting are the diverse phenotypes resulting from targeted deletion of a fraction of known RGS proteins/splice forms. Given that RGS proteins play a critical role in GPCR signaling whose dysregulation underlies many human diseases, future studies employing new genome editing tools should yield incredibly exciting insights into the physiological and pathological roles of other RGS proteins.

The enthusiasm with which the contributors to this project responded to my solicitation was very gratifying. To those authors and coauthors recruited in writing, I thank you for your time and effort in preparation of your outstanding contributions. I am particularly grateful to Adele Stewart for helping me conceive and contribute to this volume. I thank all of the authors for your friendly way in responding to my minor editorial suggestions. This made my job a pleasant and rewarding experience. Special thanks to P. Michael Conn, friend and Chief Editor of the *Progress in Molecular Biology and Translational Science* series, for deciding to choose this volume on RGS proteins and for providing me the opportunity to become involved. Finally, it has been wonderful to work with the colleagues at Elsevier, especially Roshmi Joy and Helene Kabes. Their support and help in moving the project along is sincerely appreciated.

Rory A. Fisher

CHAPTER ONE

Introduction: G Protein-coupled Receptors and RGS Proteins

Adele Stewart[2], Rory A. Fisher[1]
Department of Pharmacology, University of Iowa Carver College of Medicine, Iowa City, Iowa, USA
[1]Corresponding author: e-mail address: rory-fisher@uiowa.edu

Contents

Abstract

Here, we provide an overview of the role of regulator of G protein-signaling (RGS) proteins in signaling by G protein-coupled receptors (GPCRs), the latter of which represent the largest class of cell surface receptors in humans responsible for transducing diverse extracellular signals into the intracellular environment. Given that GPCRs regulate virtually every known physiological process, it is unsurprising that their dysregulation plays a causative role in many human diseases and they are targets of 40–50% of currently marketed pharmaceuticals. Activated GPCRs function as GTPase exchange factors for Gα subunits of heterotrimeric G proteins, promoting the formation of Gα-GTP and dissociated Gβγ subunits that regulate diverse effectors including enzymes, ion channels, and protein kinases. Termination of signaling is mediated by the intrinsic GTPase activity of Gα subunits leading to reformation of the inactive Gαβγ heterotrimer. RGS proteins determine the magnitude and duration of cellular responses initiated by many GPCRs by functioning as GTPase-accelerating proteins (GAPs) for specific Gα subunits. Twenty canonical mammalian RGS proteins, divided into four subfamilies, act as functional GAPs while almost 20 additional proteins contain nonfunctional RGS homology domains that often mediate interaction with GPCRs or Gα subunits. RGS protein biochemistry has been well elucidated *in vitro*, but the physiological functions of each RGS family member remain largely unexplored. This book summarizes recent advances employing modified model organisms that reveal RGS protein functions *in vivo*, providing evidence that RGS protein modulation of G protein signaling and GPCRs can be as important as initiation of signaling by GPCRs.

[2] Present address: Department of Pharmacology, Vanderbilt University, Nashville, Tennessee, USA

Progress in Molecular Biology and Translational Science, Volume 133
ISSN 1877-1173
http://dx.doi.org/10.1016/bs.pmbts.2015.03.002

1. GPCR PHYSIOLOGY, PATHOPHYSIOLOGY, AND PHARMACOLOGY

G protein-coupled receptors (GPCRs) represent the largest class of cell surface receptors and are responsible for transducing extracellular signals in the form of peptides, neurotransmitters, hormones, odorants, light, ions, nucleotides, or amino acids into the intracellular environment. It is now believed that the GPCR superfamily contains over 1000 genes in humans, comprising ~2% of all gene-encoding DNA.[1,2] Given the diversity of GCPR stimuli and the abundance of GPCR-encoding genes in the human genome, it is not surprising that GPCR dysregulation plays a causative role in many human maladies including cardiovascular diseases, neuropsychiatric disorders, metabolic syndromes, carcinogenesis, and viral infections.[3–6] In fact, it is estimated that 40–50% of currently marketed pharmaceuticals target GPCRs, arguably the most remunerative drug class with worldwide sales totaling $47 billion in 2003.[3]

Though new GPCR-targeted drugs are in the pharmaceutical industry pipeline,[7] a number of challenges have emerged in the development of novel therapeutics aimed at disrupting or enhancing signaling through GPCRs. In particular, for many years, a lack of high-resolution crystal structures made *in silico* bioinformatic drug screening challenging. The recently solved structure of the β_2-adrenergic receptor in complex with $G\alpha_s$[8] (amongst others) will likely facilitate such efforts in the coming years. Additional hurdles in GPCR drug development include agonist-induced receptor desensitization and tolerance; activation or inhibition of multiple GPCR effector cascades; a lack of selectivity between ligand-specific receptor subtypes; and the possibility of off-target effects due to receptor expression in multiple cells, tissues or organs in the body.[7] Though receptor targeting is ideal due to the lack of need for intracellular drug trafficking, it is now believed that GPCR effectors and regulators may also be viable drug targets and might represent a means to improve therapeutic efficacy and specificity.

2. GPCR SIGNAL TRANSDUCTION: HETEROTRIMERIC G PROTEINS

Structurally, GPCRs are characterized by seven membrane-spanning alpha helices with an extracellular N-terminal tail, often, but not exclusively, involved in ligand binding, and intracellular loops and a C-terminus involved in guanine-nucleotide regulatory protein (G protein) coupling and receptor regulation. Ligand binding is believed to induce a conformational change in the receptor that promotes G protein association.[9] Activated receptors function as guanine nucleotide exchange factors (GEFs) for the α subunit of the heterotrimeric G protein complex. Gα will then transition from its inactive guanosine diphosphate (GDP)-bound form to the active guanosine triphosphate (GTP)-bound monomer, dissociating from the Gβγ dimer (Fig. 1). There are four families of Gα subunits in mammals ($G\alpha_s$, $G\alpha_i$, $G\alpha_q$, and $G\alpha_{12/13}$), which differ in their specific effector coupling, downstream signaling, and net cellular response. GPCR coupling to Gα subunits is highly selective allowing for ligand-specific modulation of downstream signaling in cells. Gα subunits contain two characterized functional domains: a GTP-binding cassette homologous to that found in Ras-like small GTPases and a helical insertion. GCPRs trigger a conformational change in the three flexible "switch" regions of the GTP-binding domain. The helical insertion, conversely, is unique to heterotrimeric G proteins and functions to sequester the guanine nucleotide in the GTP-binding domain. Nucleotide dissociation requires displacement of this structure, a process facilitated by active GPCRs.[10,11] Both GTP-bound Gα and Gβγ activate effector molecules, which include enzymes, ion channels, and protein kinases.[3] Deactivation of G-protein signaling occurs by the

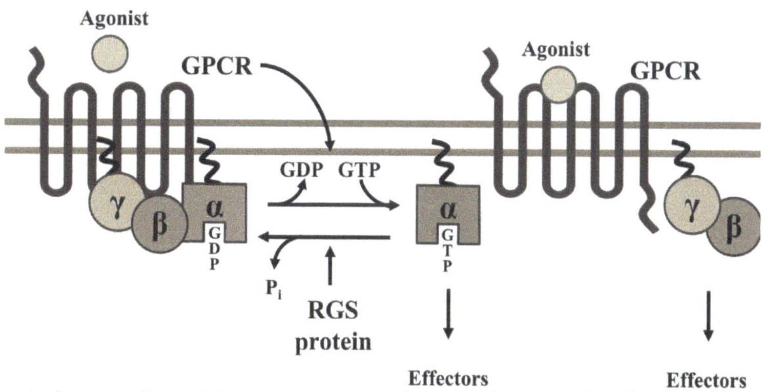

Figure 1 Canonical regulation of GPCR signaling by RGS proteins. Agonist binding to GPCRs induces a conformation change that facilitates the exchange of GDP for GTP on the α subunit of the heterotrimeric complex. Both GTP-bound Gα in the active form and the released Gβγ dimer can then go on to stimulate a number of downstream effectors. RGS proteins are GAPs for Gα, which function to terminate signaling through GPCRs by accelerating the intrinsic GTPase activity of Gα and promoting reassociation of the heterotrimeric complex with the receptor at the cell membrane. (See the color plate.)

intrinsic hydrolysis of GTP to GDP by the Gα subunit, which occurs at a rate that varies among the G-protein subfamilies.[12]

Five genes encode Gβ subunits and twelve genes encode the varying Gγ isoforms resulting in an impressive diversity of possible dimeric Gβγ complexes.[13] Gβ and Gγ subunits form obligate heterodimers *in vivo* as Gβ requires Gγ for proper protein folding.[14] Gγ proteins have a simple structure containing two α-helices joined by a linker loop, which form a coiled-coil interaction with the N-terminal α-helix of Gβ.[15] The remainder of the Gβ subunit consists of a β-propeller motif composed of tryptophan-aspartic acid (WD) repeats forming arrangements of antiparallel β sheets. Crystal structures of effector-bound Gβγ complexes have revealed that this β-propeller structure is intimately involved in effector coupling.[16,17] Unsurprisingly, this effector-binding site largely overlaps with the region responsible for interaction between Gβγ dimers and the switch II region of Gα, which explains the lack of Gβγ signaling when sequestered in the heterotrimeric G protein complex.[12] It is known that some Gβ and Gγ subunits preferentially interact[18–20] leading to the supposition that there may be some selectivity in Gβγ dimer receptor/G protein coupling and effector activation. Indeed, studies in individual Gβ and Gγ knockout models have revealed unique phenotypic consequences for loss of specific subunits implying that these proteins are not as interchangeable as was originally believed.[21]

3. G PROTEIN REGULATION

Regulation of GPCRs is complex with multiple layers of interconnected signaling pathways activated upon receptor simulation that feedback to impact receptor function. The best characterized GPCR regulatory mechanisms are mediated by G protein-coupled receptor kinases (GRKs), arrestins, and regulator of G protein-signaling (RGS) proteins. The Gβγ dimer facilitates membrane targeting of GRKs resulting in GRK-mediated GPCR phosphorylation. This modification recruits β-arrestins, which sterically hinder further G-protein coupling to the receptor.[22] Though their role in GPCR desensitization has been well characterized, it is now appreciated that arrestins are multifunctional scaffolds involved in numerous aspects of GCPR signal transduction.[23]

In the late 1980s, a discrepancy was noted between the biochemical GTPase activity of Gα subunits and the turnoff rate for the cellular response to endogenous GPCR ligands. The so-called "missing link" was discovered in the founding members of the RGS protein family identified in yeast[24] and

Caenorhabditis elegans,[25] which shared sequence homology with a larger group of mammalian proteins. The prototypic role of RGS proteins is negative regulation of G protein signaling through acceleration of GTP hydrolysis by Gα. In so doing, RGS proteins promote reassociation of Gα and Gβγ subunits with the receptor at the cell membrane and terminate signaling of both Gα and Gβγ to downstream effectors (Fig. 1). In this way, RGS proteins determine the magnitude and duration of the cellular response to GPCR stimulation.[26,27]

4. RGS PROTEINS

Twenty canonical mammalian RGS proteins, divided into four subfamilies based on sequence homology and the presence and nature of additional non-RGS domains, act as functional GTPase accelerating proteins (GAPs) for $G\alpha_{i/o}$, $G\alpha_{q/11}$ or both. Almost 20 additional proteins contain nonfunctional RGS homology domains that often mediate interaction with GPCRs or Gα subunits (Table 1). Functional RGS proteins share a conserved core interface that mediates the interaction with Gα subunits. Adjacent modulatory residues determine G protein specificity or lack thereof.[33] The mechanism of RGS protein-mediated acceleration of GTP hydrolysis by Gα has been inferred from crystal structures of the RGS protein–Gα complex.[34] Because the trio of conserved Gα residues necessary for GTP hydrolysis is sufficient for this activity, RGS protein are not traditional enzymes and, instead, stabilize the transition state conformation lowering the free energy required to activate the hydrolysis reaction.[34,35] RGS protein biochemistry has been well elucidated *in vitro*, but the physiological functions of each RGS family member remain largely unexplored.

Historically, a lack of specific antibodies with corresponding genetic knockout controls has made detection of endogenous RGS proteins difficult *in vivo*, making investigations of the physiological significance of RGS proteins even more challenging. Because most tissues express multiple RGS transcripts encoding proteins that would be capable of acting as functional GAPs for the same Gα subunits, one major challenge in investigating RGS protein function in living animals is the potential for functional redundancy and compensatory changes in RGS protein expression that result from loss of a single protein. Indeed, the phenotypes of single RGS protein knockouts are usually modest in the absence of a physiological or pathophysiological stimulus. Combinatorial knockout of two or more RGS protein in order to investigate the net importance of RGS protein function in a

Table 1 RGS Protein Superfamily

Family	Member	Gα GAP Activity	Additional Structural Motifs and Domains
A/RZ	RGS17 (RGSZ2)	$G\alpha_{i/o}$ and $G\alpha_z$	Cys
	RGS19 (GAIP)	$G\alpha_{i/o}$, $G\alpha_{q/11}$, and $G\alpha_z$	Cys
	RGS20 (RGSZ1)	$G\alpha_z$	Cys
B/R4	RGS1	$G\alpha_{i/o}$ and $G\alpha_{q/11}$	AH
	RGS2	$G\alpha_q$	AH
	RGS3	$G\alpha_{i/o}$ and $G\alpha_{q/11}$	AH
	RGS4	$G\alpha_{i/o}$ and $G\alpha_{q/11}$	AH
	RGS5	$G\alpha_{i/o}$ and $G\alpha_{q/11}$	AH
	RGS8	$G\alpha_{i/o}$ and $G\alpha_{q/11}$	AH
	RGS13	$G\alpha_{i/o}$ and $G\alpha_{q/11}$	AH
	RGS16	$G\alpha_{i/o}$ and $G\alpha_{q/11}$	AH
	RGS18	$G\alpha_{i/o}$ and $G\alpha_{q/11}$	AH
	RGS21	ND	
C/R7	RGS6	$G\alpha_{i/o}$	DEP/DHEX, GGL
	RGS7	$G\alpha_{i/o}$	DEP/DHEX, GGL
	RGS9	$G\alpha_{i/o}$	DEP/DHEX, GGL
	RGS11	$G\alpha_{i/o}$	DEP/DHEX, GGL
D/R12	RGS10	$G\alpha_{i/o}$ and $G\alpha_{q/11}$	
	RGS12	$G\alpha_{i/o}$	PDZ, PTB, RBD (2), GoLoco
	RGS14	$G\alpha_{i/o}$	RBD (2), GoLoco
E/RA	Axin	N/A	CC, DAX, GSK3β BD. β-catenin BD
	Axin 2	N/A	CC, DAX
F/GEF	p115-RhoGEF	N/A	CC, DH, PH
	PDZ-RhoGEF	N/A	CC, DH, PH, PDZ
	LARG	N/A	CC, DH, PH, PDZ

Table 1 RGS Protein Superfamily—cont'd

Family	Member	Gα GAP Activity	Additional Structural Motifs and Domains
G/GRK	GRK1	N/A	S/T kinase
	GRK2	N/A	S/T kinase, PH, CC
	GRK3	N/A	S/T kinase, PH
	GRK4-7	N/A	S/T kinase
H/SNX	SNX13	N/A	TMD (2), PXA, PX, CC (2)
	SNX14	N/A	TMD (2), PXA, PX
	SNX25	N/A	PXA, PX, CC
Other	RGS22	$G\alpha_{12/13}$ and $G\alpha_{q/11}$	
	D-AKAP2	N/A	PKA BD

This table lists proteins with functional RGS domains or nonfunctional RGS homology domains. RGS proteins are grouped into subclasses based on sequence homology, GAP specificity, and the presence of additional functional domains or structural motifs.[28–32]

Note: Abbreviations used are AH, amphiphatic helix; β-catenin BD, β-catenin binding domain; CC, coiled coil motif; Cys, cysteine string; DAX, domain present in disheveled and axin; DEP, disheveled, EGL-10, pleckstrin homology domain; DH, Dbl homology domain; DHEX, DEP helical extension; GGL, Gγ subunit-like domain; GoLoco, G protein regulatory motif; GSK3β BD, GSK3β-binding domain; N/A, not applicable; ND, not determined; PDZ, domain present in PSD-95, Dlg, and ZO-1/2; PH, pleckstrin homology domain; PKA BD, PKA-binding domain; PTB, phosphotyrosine-binding domain; PC, PhoX homologous domain; PXA, PX-associated domain; RBD, Raf-like Ras binding domain; S/T kinase, serine/threonine kinase domain; TMD, transmembrane domain.

particular disease or physiological process is a technical and financial nightmare.[36]

To circumvent these issues, a series of transgenic mice were developed that express knock-in alleles of RGS-insensitive Gα mutants. In place of the endogenous protein, these mice instead express Gα with a point mutation (G184S in $G\alpha_{i2}$) in the switch I region that blocks the interaction with RGS proteins necessary for GTPase activation[37] without affecting the intrinsic GTPase activity of Gα or its ability to bind Gβγ, GPCRs, and effectors.[38] Thus these mouse models have been used to evaluate the net regulatory actions of RGS proteins on various GPCR signaling pathways *in vivo*. Studies in these animals revealed that endogenous RGS proteins play critical roles in controlling cardiovascular biology, metabolism, inflammation, anxiety and depression, and pain (Table 2).

Table 2 Reported Phenotypes of Knock-In Mice Expressing RGS-Insensitive Gα Mutants

Gα Subunit	Phenotype	Year	Reference(s)
$G\alpha_{i2}$(G148S)	Reduced viability, growth retardation, hyperactivity, hematologic abnormalities, cardiac hypertrophy	2006	39
	Enhanced parasympathetic stimulation of heart	2007	40
	Resistance to diet-induced obesity and insulin resistance	2008	41
	Potentiation of epinephrine-mediated antiepileptic actions	2009	42
	Alterations in isoflurane-induced loss of righting reflex and breathing	2009	43
	Exacerbated platelet accumulation and thrombus formation following vascular injury	2010	44
	Baseline reduction in anxiety- and depression-related behaviors	2010	45
	Protection from ischemic cardiac injury	2011	46
	Increased cardiac hypertrophy in genetic- and catecholamine-induced models of cardiomyopathy	2012	47
	Protection from endotoxemia-induced proinflammatory cytokine production	2012	48
	Deficit in neutrophil mobilization to sites of inflammation and infection, myelokathexis	2012	49
$G\alpha_o$(G148S)	Enhanced thermal analgesia in response to endogenous and exogenous opioids	2013	50

The various phenotypes of Gα(G148S) mutant knock-in mice are listed in chronological order with associated references. The phenotypes of these mice represent the functional consequence of loss of all RGS protein-mediated regulation of Gα signaling.

This book summarizes the current state of the RGS protein field, describing demonstrated RGS protein functions *in vivo* identified using genetically modified model organisms.

REFERENCES

1. Jacoby E, Bouhelal R, Gerspacher M, Seuwen K. The 7 TM G-protein-coupled receptor target family. *ChemMedChem*. 2006;1(8):761–782.

2. Insel PA, Snead A, Murray F, et al. GPCR expression in tissues and cells: are the optimal receptors being used as drug targets? *Br J Pharmacol.* 2012;165(6):1613–1616.
3. Lundstrom KH, Chiu ML. *G Protein-Coupled Receptors in Drug Discovery.* Boca Raton, FL: Taylor & Francis; 2006.
4. Salazar NC, Chen J, Rockman HA. Cardiac GPCRs: GPCR signaling in healthy and failing hearts. *Biochim Biophys Acta.* 2007;1768(4):1006–1018.
5. Lappano R, Maggiolini M. G protein-coupled receptors: novel targets for drug discovery in cancer. *Nat Rev Drug Discov.* 2011;10(1):47–60.
6. Insel PA, Tang CM, Hahntow I, Michel MC. Impact of GPCRs in clinical medicine: monogenic diseases, genetic variants and drug targets. *Biochim Biophys Acta.* 2007;1768(4):994–1005.
7. Garland SL. Are GPCRs still a source of new targets? *J Biomol Screen.* 2013;18(9):947–966.
8. Rasmussen SG, DeVree BT, Zou Y, et al. Crystal structure of the beta2 adrenergic receptor-Gs protein complex. *Nature.* 2011;477(7366):549–555.
9. Chung KY. Structural aspects of GPCR-G protein coupling. *Toxicol Res.* 2013;29(3):149–155.
10. Lambright DG, Noel JP, Hamm HE, Sigler PB. Structural determinants for activation of the alpha-subunit of a heterotrimeric G protein. *Nature.* 1994;369(6482):621–628.
11. Wall MA, Posner BA, Sprang SR. Structural basis of activity and subunit recognition in G protein heterotrimers. *Structure.* 1998;6(9):1169–1183.
12. Johnston CA, Siderovski DP. Receptor-mediated activation of heterotrimeric G-proteins: current structural insights. *Mol Pharmacol.* 2007;72(2):219–230.
13. Hamm HE. The many faces of G protein signaling. *J Biol Chem.* 1998;273(2):669–672.
14. Higgins JB, Casey PJ. *In vitro* processing of recombinant G protein gamma subunits. Requirements for assembly of an active beta gamma complex. *J Biol Chem.* 1994;269(12):9067–9073.
15. Sondek J, Bohm A, Lambright DG, Hamm HE, Sigler PB. Crystal structure of a G-protein beta gamma dimer at 2.1A resolution. *Nature.* 1996;379(6563):369–374.
16. Ford CE, Skiba NP, Bae H, et al. Molecular basis for interactions of G protein betagamma subunits with effectors. *Science.* 1998;280(5367):1271–1274.
17. Gaudet R, Bohm A, Sigler PB. Crystal structure at 2.4 angstroms resolution of the complex of transducin betagamma and its regulator, phosducin. *Cell.* 1996;87(3):577–588.
18. Lee C, Murakami T, Simonds WF. Identification of a discrete region of the G protein gamma subunit conferring selectivity in beta gamma complex formation. *J Biol Chem.* 1995;270(15):8779–8784.
19. Poon LS, Chan AS, Wong YH. Gbeta3 forms distinct dimers with specific Ggamma subunits and preferentially activates the beta3 isoform of phospholipase C. *Cell Signal.* 2009;21(5):737–744.
20. Yan K, Kalyanaraman V, Gautam N. Differential ability to form the G protein betagamma complex among members of the beta and gamma subunit families. *J Biol Chem.* 1996;271(12):7141–7146.
21. Khan SM, Sleno R, Gora S, et al. The expanding roles of Gbetagamma subunits in G protein-coupled receptor signaling and drug action. *Pharmacol Rev.* 2013; 65(2):545–577.
22. Krupnick JG, Benovic JL. The role of receptor kinases and arrestins in G protein-coupled receptor regulation. *Annu Rev Pharmacol Toxicol.* 1998;38:289–319.
23. Patel PA, Tilley DG, Rockman HA. Physiologic and cardiac roles of beta-arrestins. *J Mol Cell Cardio.* 2009;46(3):300–308.
24. Dohlman HG, Apaniesk D, Chen Y, Song J, Nusskern D. Inhibition of G-protein signaling by dominant gain-of-function mutations in Sst2p, a pheromone desensitization factor in Saccharomyces cerevisiae. *Mol Cell Biol.* 1995;15(7):3635–3643.

25. Koelle MR, Horvitz HR. EGL-10 regulates G protein signaling in the C. elegans nervous system and shares a conserved domain with many mammalian proteins. *Cell.* 1996;84(1):115–125.
26. Berman DM, Wilkie TM, Gilman AG. GAIP and RGS4 are GTPase-activating proteins for the Gi subfamily of G protein alpha subunits. *Cell.* 1996;86(3):445–452.
27. Ross EM, Wilkie TM. GTPase-activating proteins for heterotrimeric G proteins: regulators of G protein signaling (RGS) and RGS-like proteins. *Annu Rev Biochem.* 2000;69:795–827 (Review).
28. Hurst JH, Hooks SB. Regulator of G-protein signaling (RGS) proteins in cancer biology. *Biochem Pharmacol.* 2009;78(10):1289–1297.
29. Willars GB. Mammalian RGS, proteins: multifunctional regulators of cellular signalling. *Semin Cell Dev Biol.* 2006;17(3):363–376.
30. Tesmer JJ. Structure and function of regulator of G protein signaling homology domains. *Prog Mol Biol Transl Sci.* 2009;86:75–113.
31. Sjogren B, Blazer LL, Neubig RR. Regulators of G protein signaling proteins as targets for drug discovery. *Prog Mol Biol Transl Sci.* 2010;91:81–119.
32. Zhang P, Mende U. Regulators of G-protein signaling in the heart and their potential as therapeutic targets. *Circ Res.* 2011;109(3):320–333.
33. Kosloff M, Travis AM, Bosch DE, Siderovski DP, Arshavsky VY. Integrating energy calculations with functional assays to decipher the specificity of G protein-RGS protein interactions. *Nat Struct Mol Biol.* 2011;18(7):846–853.
34. Tesmer JJ, Berman DM, Gilman AG, Sprang SR. Structure of RGS4 bound to AlF4-activated G(i alpha1): stabilization of the transition state for GTP hydrolysis. *Cell.* 1997;89(2):251–261.
35. Berman DM, Kozasa T, Gilman AG. The GTPase-activating protein RGS4 stabilizes the transition state for nucleotide hydrolysis. *J Biol Chem.* 1996;271(44):27209–27212.
36. Kaur K, Kehrl JM, Charbeneau RA, Neubig RR. RGS-insensitive Galpha subunits: probes of Galpha subtype-selective signaling and physiological functions of RGS proteins. *Methods Mol Biol.* 2011;756:75–98.
37. Lan KL, Sarvazyan NA, Taussig R, et al. A point mutation in Galphao and Galphai1 blocks interaction with regulator of G protein signaling proteins. *J Biol Chem.* 1998;273(21):12794–12797.
38. Fu Y, Zhong H, Nanamori M, et al. RGS-insensitive G-protein mutations to study the role of endogenous RGS proteins. *Methods Enzymol.* 2004;389:229–243.
39. Huang X, Fu Y, Charbeneau RA, et al. Pleiotropic phenotype of a genomic knock-in of an RGS-insensitive G184S Gnai2 allele. *Mol Cell Biol.* 2006;26(18):6870–6879.
40. Fu Y, Huang X, Piao L, Lopatin AN, Neubig RR. Endogenous RGS proteins modulate SA and AV nodal functions in isolated heart: implications for sick sinus syndrome and AV block. *Am J Physiol Heart Circ Physiol.* 2007;292(5):H2532–2539.
41. Huang X, Charbeneau RA, Fu Y, et al. Resistance to diet-induced obesity and improved insulin sensitivity in mice with a regulator of G protein signaling-insensitive G184S Gnai2 allele. *Diabetes.* 2008;57(1):77–85.
42. Goldenstein BL, Nelson BW, Xu K, et al. Regulator of G protein signaling protein suppression of Galphao protein-mediated alpha2A adrenergic receptor inhibition of mouse hippocampal CA3 epileptiform activity. *Mol Pharmacol.* 2009;75(5):1222–1230.
43. Icaza EE, Huang X, Fu Y, Neubig RR, Baghdoyan HA, Lydic R. Isoflurane-induced-changes in righting response and breathing are modulated by RGS proteins. *Anesth Analg.* 2009;109(5):1500–1505.
44. Signarvic RS, Cierniewska A, Stalker TJ, et al. RGS/Gi2alpha interactions modulate platelet accumulation and thrombus formation at sites of vascular injury. *Blood.* 2010;116(26):6092–6100.

45. Talbot JN, Jutkiewicz EM, Graves SM, et al. RGS inhibition at G(alpha)i2 selectively potentiates 5-HT1A-mediated antidepressant effects. *Proc Natl Acad Sci U S A.* 2010;107(24):11086–11091.
46. Waterson RE, Thompson CG, Mabe NW, et al. Galpha(i2)-mediated protection from ischaemic injury is modulated by endogenous RGS proteins in the mouse heart. *Cardiovasc Res.* 2011;91(1):45–52.
47. Kaur K, Parra S, Chen R, et al. Galphai2 signaling: friend or foe in cardiac injury and heart failure? *Naunyn Schmiedebergs Arch Pharmacol.* 2012;385(5):443–453.
48. Li P, Neubig RR, Zingarelli B, et al. Toll-like receptor-induced inflammatory cytokines are suppressed by gain of function or overexpression of Galpha(i2) protein. *Inflammation.* 2012;35(5):1611–1617.
49. Cho H, Kamenyeva O, Yung S, et al. The loss of RGS protein-Galpha(i2) interactions results in markedly impaired mouse neutrophil trafficking to inflammatory sites. *Mol Cell Biol.* 2012;32(22):4561–4571.
50. Lamberts JT, Smith CE, Li MH, Ingram SL, Neubig RR, Traynor JR. Differential control of opioid antinociception to thermal stimuli in a knock-in mouse expressing regulator of G-protein signaling-insensitive Galphao protein. *J Neurosci.* 2013; 33(10):4369–4377.

CHAPTER TWO

RGS-Insensitive G Proteins as *In Vivo* Probes of RGS Function

Richard R. Neubig[1]
Department of Pharmacology & Toxicology, Michigan State University, East Lansing, Michigan, USA
[1]Corresponding author: e-mail address: RNeubig@msu.edu

Contents

Abstract

Guanine nucleotide-binding proteins of the inhibitory ($G_{i/o}$) class play critical physiological roles and the receptors that activate them are important therapeutic targets (e.g., mu opioid, serotonin $5HT_{1a}$, etc.). $G_{i/o}$ proteins are negatively regulated by regulator of G protein signaling (RGS) proteins. The redundant actions of the 20 different RGS family members have made it difficult to establish their overall physiological role. A unique G protein mutation (G184S in $G\alpha_{i/o}$) prevents RGS binding to the Gα subunit and blocks all RGS action at that particular Gα subunit. The robust phenotypes of mice expressing these RGS-insensitive (RGSi) mutant G proteins illustrate the profound action of RGS proteins in cardiovascular, metabolic, and central nervous system functions. Specifically, the enhanced $G\alpha_{i2}$ signaling through the RGSi $G\alpha_{i2}^{G184S}$ mutant knock-in mice shows protection against cardiac ischemia/reperfusion injury and potentiation of serotonin-mediated antidepressant actions. In contrast, the RGSi $G\alpha_o$ mutant knock-in produces enhanced mu-opioid receptor-mediated analgesia but also a seizure phenotype. These

Progress in Molecular Biology and Translational Science, Volume 133
ISSN 1877-1173
http://dx.doi.org/10.1016/bs.pmbts.2015.04.010

genetic models provide novel insights into potential therapeutic strategies related to RGS protein inhibitors and/or G protein subtype-biased agonists at particular GPCRs.

1. INTRODUCTION

There are four major G protein families, G_s, G_i, G_q, and G_{12}.[1] This chapter will focus on regulator of G protein signaling (RGS) protein control of the G_i (or $G_{i/o}$) family of G proteins. Along with the G_q family, the $G_{i/o}$ family is a definitive target of RGS protein regulation.[2,3] The $G_{i/o}$ family includes G_{i1}, G_{i2}, and G_{i3}, G_o, G_z, and the sensory G proteins, G_{tr}, G_{tc}, and G_{gust}. They all associate with a Gβγ subunit like all heterotrimeric G proteins. In contrast to the G_s, G_q, and G_{12} families, however, the signaling mechanism of the G_i-family proteins is usually mediated through the released Gβγ subunits. Signals downstream of $G_{i/o}$ include modulation of adenylyl cyclase (AC) which for most AC subtypes is inhibition,[4] activation of G protein-coupled inwardly rectifying potassium channels (GIRK), inhibition of N-, and P/Q-type Ca^{++} channels,[5,6] activation of PLC, activation of PI-3-kinase (β and γ isoforms), and activation of ERK MAPK by diverse and often indirect mechanisms.[1] In some cases (e.g., $G\alpha_o$-mediated inhibition of Type 1 and $G\alpha_i$-mediated inhibition of Type 5 and 6 AC), the alpha subunit mediates the signal. Whether the Gα or the Gβγ subunit mediates the signal, the regulation by both receptors and RGS proteins is largely similar. Receptors increase the amount of the active subunits, Gα-GTP and free Gβγ, while RGS proteins reduce them.

A common feature of all the G_i-family G proteins, except G_z, is their sensitivity to pertussis toxin.[1] Consequently, any physiological signal in non-sensory tissues that is sensitive to pertussis toxin is almost certainly mediated by $G_{i1,\ 2,\ \mathrm{or}\ 3}$, or G_o. Thus, it is relatively common to see a function attributed to $G_{i/o}$ without any further distinction among the four. One point made in this chapter is the ability of the mutant $G_{i/o}$ subunits to provide information about $G_{i/o}$ subtype-specific functions.

2. GENETIC MODELS OF THE ROLE OF RGS IN PHYSIOLOGY AND PATHOPHYSIOLOGY

The RGS proteins were initially discovered through genetic studies in model organisms, such as yeast (*Saccharomyces cerevisiae*) and worms (*Caenorhabditis elegans*).[7–10] Similarly, multiple genetic studies of RGS proteins in mice[11–14] and a few rare RGS mutations in humans[15,16] have revealed much information about the physiological functions of RGS proteins. Such studies are very valuable in assessing the potential roles of RGS proteins as novel drug targets.[2,17–19] The pros and cons of RGS knock-out animal models and a comparison of alternative approaches to define the functional importance of RGS proteins are outlined in Table 1.

Table 1 Distinct Approaches to Evaluating Physiological Functions of RGS Proteins

	Normal Function	RGS Knockout	Chemical RGS Inhibitors	RGS-Insensitive Gα Subunit Mutants
Expected effects	Reduced signal	Enhanced signal	Enhanced signal	Enhanced signal
Specificity	Controlled by level and location of expression of *multiple* RGS proteins	Controlled by level and location of *specific* RGS protein expression	Controlled by *compound specificity* and level and location of various RGS proteins	Controlled by level and location of *G protein* and its modulation by all RGS proteins
Advantages		Targeted to one RGS protein; redundancy with multiple RGS proteins blunts phenotypes	Controlled onset and duration; therapeutically relevant	Avoids problem of redundancy of RGS proteins; provides insights to G protein role (e.g., $G\alpha_{i1}$ vs. $G\alpha_{i2}$); robust phenotypes
Limitations		Developmental effects	Off-target effects on both other RGS proteins and non-RGS functions	Low viability of mutants; does not show which RGS is relevant
Alternative strategies to overcome limitations		Conditional knockout	Compound optimization	Conditional expression

There are a number of experimental methods to assess function of RGS proteins. The three most common methods are outlined here with their various pros and cons.

In addition to traditional and conditional RGS gene knockouts which have been reviewed previously,[14] an alternative, using unique G protein gain-of-function mutations was made possible by work in *S. cerevisiae*. Dohlman and colleagues[20] undertook a yeast genetic screen for mutations that phenocopy the loss of the yeast RGS protein Sst2, which results in enhanced G protein function. Using a genome-wide mutagenesis screen, they found a mutation in the Gα subunit Gpa1 (G302S) in yeast with the same phenotype (enhanced sensitivity to pheromone signaling) that the *sst2* mutant yeast had. The homologous mutation in mammalian Gαq enhanced its function in cell-based studies.[20] The related mutation G18451 in mammalian $G\alpha_{i1}$ and $G\alpha_o$ proteins completely blocked the binding of multiple RGS proteins in biochemical studies and also blocked the RGS protein's GTPase accelerating protein activity.[21]

3. RGS KNOCKOUTS VERSUS RGS-INSENSITIVE Gα KNOCK-IN MODELS

These RGS-insensitive (RGSi) Gα subunits provide an alternative to RGS protein knockouts in assessing the physiological roles of RGS proteins (Fig. 1). Table 1 compares and contrasts the information obtained from knockouts versus the RGSi approach. In brief, the RGSi approach generally

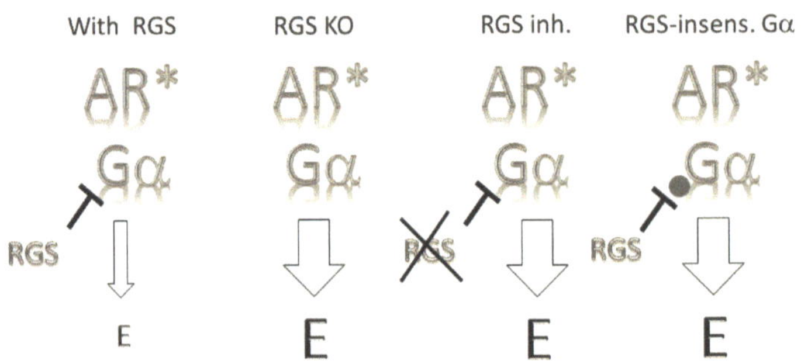

Figure 1 Comparison of distinct approaches to disrupt RGS protein functions. Agonist-activated receptors AR* induce G protein activation to produce functional effects (E). In the presence of an active RGS protein, the effect is smaller (smaller E) than in its absence (second model). RGS proteins can be knocked out themselves disrupting the action of only that RGS protein and revealing its specific actions. Redundancy with multiple RGS proteins present in the cell (cardiac cells express >10 different RGS protein mRNAs) often lead to modest effects. In contrast, the G184S mutant of the Gα subunit (last model) prevents the actions of all RGS proteins. As noted in Table 1, this provides a more robust idea of the role of RGS proteins overall but the specificity of the effects is that of the Gα subunit and not of any individual RGS protein. (See the color plate.)

provides a more robust phenotype since redundancy among the 20 different RGS proteins is eliminated. The G184S mutation in a $G\alpha_i$ or $G\alpha_o$ protein prevents the binding of any RGS protein to the $G\alpha$ subunit thus eliminating all RGS function at that specific $G\alpha$ subunit subtype. Also, given the gain-of-function effect of the G184S mutants, even the heterozygous mice are expected to have a significant increase in signaling (e.g., 46-fold for the heterozygous RGSi mutant vs. twofold or less for a heterozygous RGS knockout; see Kehrl[22]). Also, the RGSi $G_{i/o}$ protein mutation, when knocked into the endogenous genomic locus, reveals physiologic functions unique to that particular $G\alpha$ subtype. This will be governed both by the tissue-specific expression of that $G\alpha$ subunit and also its subcellular localization or any specific signal mechanisms. As noted below, the phenotypes of the $G\alpha_{i2}$ and $G\alpha_o$ G184S mutant knock-in mice are distinctly different confirming different physiological roles for the two proteins (Table 2).

4. PHENOTYPES OF $G\alpha_{i2}$ G184S MUTANT KNOCK-IN MICE

4.1 Signaling

As expected, G_i signaling in embryo fibroblasts (MEFs) derived from $G\alpha_{i2}^{+/G184S}$ heterozygotes and $G\alpha_{i2}^{G184S/G184S}$ homozygotes was enhanced. Inhibition of cAMP accumulation and activation of PI-3-kinase by lysophosphatidic acid showed greater potency or efficacy in both heterozygous and homozygous mutant MEFs.[23]

4.2 Heart

Signaling by G_i proteins has been implicated in a variety of cardiac functions, as has its regulation by RGS proteins. Inhibitory effects on heart rate, conductance, and cardiac contractility have all been observed. There is a role for both inhibition of AC and activation of GIRK currents.[36] The slowing of heart rate by vagal release of acetylcholine is mediated by M2 muscarinic receptors activating a G_i-family G protein to stimulate the GIRK channels via the $G\beta\gamma$ subunit.[37–39] Also, the kinetics of onset and offset of the muscarinic/GIRK mechanism was shown to depend on RGS proteins.[40]

4.2.1 Heart Rate

In MEFs differentiated into spontaneously contracting "atrial–nodal" type cells, the $G\alpha_{i2}^{G184S}$ knock-in mutants showed a greater than sixfold increase in potency of carbachol to reduce the beating rate.[41] Interestingly, the

Table 2 Reported Phenotypes of Knock-In Mice Expressing RGS-Insensitive Gα Mutants

Gα Subunit	Phenotype	Year	References
$G\alpha_{i2}$(G148S)	Reduced viability, growth retardation, hyperactivity, hematologic abnormalities, and cardiac hypertrophy	2006	23
	Enhanced parasympathetic stimulation of heart	2007	24
	Resistance to diet-induced obesity and insulin resistance	2008	25
	Alterations in isoflurane-induced loss of righting reflex and breathing	2009	26
	Exacerbated platelet accumulation and thrombus formation following vascular injury	2010	27
	Baseline reduction in anxiety- and depression-related behaviors	2010	28
	Protection from cardiac ischemia–reperfusion injury	2011	29
	Increased cardiac hypertrophy in genetic- and catecholamine-induced models of cardiomyopathy	2012	30
	Protection from endotoxemia-induced proinflammatory cytokine production	2012	31
Conditional $G\alpha_{i2}$(G148S)	Protection from cardiac ischemia–reperfusion injury	2014	32
$G\alpha_o$(G148S)	Potentiation of epinephrine-mediated antiepileptic actions	2009	33
	Enhanced thermal analgesia in response to endogenous and exogenous opioids	2013	34
	Spontaneous adult lethality and enhanced responsiveness to kindling models of epilepsy (potential model of human EIEE17)	2014	35

The various phenotypes of RGS-insensitive $G\alpha_{i/o}$ (G148S) mutant knock-in mice are listed in chronological order with associated references. The phenotypes of these mice represent the functional consequence of loss of all RGS protein-mediated regulation of signaling by the specific G protein (i.e., heterotrimer containing $G\alpha_{i2}$ or $G\alpha_o$).

$G\alpha_o^{G184S}$ mutant had a less dramatic effect on muscarinic control of beating rate but showed a greater enhancement than did the $G\alpha_{i2}$ mutants for adenosine A1 effects on beating rate.[41] The differential effect of $G\alpha_{i2}$ mutants to enhance muscarinic versus adenosine receptor-induced bradycardia was also seen *in vivo* in homozygous $G\alpha_{i2}^{G184S/G184S}$ mutant mice.[41] This confirms a role for $G\alpha_{i2}$ and RGS proteins in heart rate control and indicates that different receptors appear to utilize distinct G proteins for the same functional output. This raises very interesting questions about mechanisms of specificity and the potential for therapeutic targeting.

Similar effects of the RGSi mutant $G\alpha_{i2}$ were seen on cardiac conduction in Langendorf preparations.[24] The potency of carbachol to slow the beating rate of isoproterenol-stimulated hearts was enhanced in $G\alpha_{i2}$ RGSi mutant mice. Also, the onset of third-degree AV block occurred at much lower concentrations of carbachol. This could have important implications for cardiac arrhythmias. Alterations in RGS function or mutations in $G\alpha_{i2}$ itself could induce arrhythmias.

4.2.2 Contractility

Ventricular function, including cardiac contractility, is also regulated by inhibitory G protein mechanisms, which are controlled by RGS proteins. The negative inotropic effect of carbachol on isolated cardiac myocytes from $G\alpha_{i2}^{G184S/G184S}$ mutant mice was enhanced.[42] There was markedly increased potency and a modest increase in efficacy of carbachol to reduce isoproterenol-induced fiber shortening (Fig. 2A).

4.2.3 Ischemia/Reperfusion Injury

In addition to contractility, $G\alpha_{i2}$ has also been implicated in control of myocyte injury and apoptosis.[43,44] $G\alpha_{i2}$ knock-out mice show earlier death in the face of cardiac overexpression of beta adrenergic receptors.[43] Consistent with this role, the $G\alpha_{i2}{}^{G184S}$ mutant mice show enhanced cardioprotection *in vivo*.[42] Both $G\alpha_{i2}^{+/G184S}$ heterozygotes and $G\alpha_{i2}^{G184S/G184S}$ homozygotes show reduced infarct size and enhanced functional recovery after an *in vitro* ischemia/reperfusion insult (Fig. 2B; Ref. 42). The simplest interpretation of these results is that one or more endogenously expressed RGS proteins reduces $G\alpha_{i2}$ signaling and enhances the injury in the ischemia/reperfusion (I/R) models.

This could be due to either developmental alterations or short-term signaling mechanisms. Using a conditional $G\alpha_{i2}$ G184S mutant allele that can be induced in the presence of the *cre* recombinase, Parra *et al.* found that the I/R protection was still seen.[32] This indicates that short-term RGS inhibition may be sufficient to provide improved outcomes after cardiac ischemia and reperfusion. To date, the specific receptor and RGS protein mediating these effects have not been identified. If a specific RGS protein was found to underlie this effect, then pharmacological

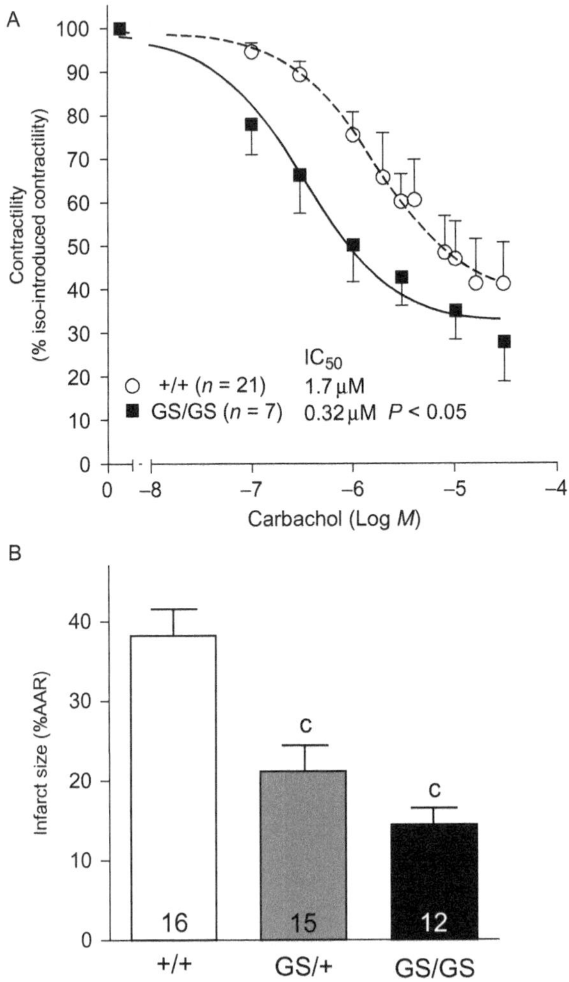

Figure 2 Enhanced negative inotropy to carbachol and protection from ischemic reperfusion injury. C57BL/6J mice with either wild-type $G\alpha_{i2}$ (+/+) or heterozygous (GS/+) or homozygous (GS/GS) knock-in G184S mutant $G\alpha_{i2}$ were assessed for (A) cardiac function and (B) ischemia/reperfusion injury. (A) Ventricular myocytes were isolated from WT or mutant mice and contractile function monitored in the presence of 100 nM isoproterenol using an IonOptix system. Variable concentrations of carbachol cause a muscarinic receptor-dependent negative inotropic effect. Signaling was significantly potentiated as carbachol had a fivefold greater potency and slightly greater efficacy on homozygous mutant myocytes to suppress the isoproterenol effect. (B) Hearts from WT and mutant mice were perfused on a Langendorf apparatus and subjected to an ischemia reperfusion protocol as described. The infarct size as a fraction of the area at risk was markedly reduced in the mutant mice (C) ($P < 0.05$ for GS/+ and GS/GS) compared to wild type. *Reproduced from Waterson* et al.[42] *with permission.*

targeting of that RGS protein could be a novel therapeutic approach in myocardial infarction. Angioplasty or stent surgeries would be obvious clinical correlates to these I/R experimental models.

4.2.4 Heart Failure/Fibrosis

While the enhanced protection against I/R injury in the $G\alpha_{i2}^{G184S}$ mutant mice could be beneficial if it could be replicated pharmacologically, a more complex effect was seen in heart failure models. As noted above, $G\alpha_{i2}$ knock-out mice show worsened heart failure in the face of beta receptor overexpression.[43] Despite this, gain of $G\alpha_{i2}$ function in the RGSi mutants did not project against hypertrophy or death in two distinct heart failure models.[30] This may have resulted from actions in nonmyocardial cells *in vivo*. Cardiac fibroblasts have been implicated as playing an adverse role in cardiac remodeling.[45] Kaur *et al.*[30] found that fibroblasts from $G\alpha_{i2}^{+/G184S}$ mutant mice were hyperproliferative and had a G_i-dependent enhancement of ERK MAPK kinase activation. This action in fibroblasts may have counteracted any potentially beneficial effect in the cardiomyocytes. Thus any attempt to produce cardioprotection by inhibiting RGS function would need to carefully define the correct RGS protein target to avoid complications of worsened heart failure.

4.2.5 Inflammation and Immunity

$G\alpha_{i2}$ also plays an important role in lymphocyte, neutrophil, and macrophage development, trafficking and activation.[46–49] RGSi $G\alpha_{i2}$ mutant mice also show altered trafficking of B lymphocytes and neutrophils.[50,51] The observation that either knock-out or gain-of-function mutations in $G\alpha_{i2}$ results in impaired migration is intriguing. This is likely due to the requirement for rapid onset and offset of signals in the context of chemotactic behaviors. One RGS protein that plays a key role in the lymphocyte actions is RGS1.[52,53] Interestingly, $G\alpha_{i2}$ appears to mediate anti-inflammatory responses as demonstrated for Toll-like receptors using RGSi mutant mice[31] and for T-cell receptors and dietary antigens using $G\alpha_{i2}^{-/-}$ mice.[54,55] Understanding more fully which RGS proteins may be suppressing these anti-inflammatory signals could prove valuable.

4.3 Central Nervous System and Depression

The specificity of different $G_{i/o}$ subtypes in mediating various physiological signals has remained incompletely characterized. One mechanism with very complex pharmacology and regulation is the role of serotonin (5HT) in depression. Enhanced 5HT signaling clearly underlies the action of selective serotonin reuptake inhibitors (SSRIs) but there is controversy as to which of the many 5HT receptors is responsible for the therapeutic effects but the $G_{i/o}$-coupled 5HT1a receptor is a prime candidate. Due to its combined pre- and postsynaptic actions, the 5HT1a receptor can mediate both enhanced and suppressed behavioral responses.[56] Interestingly, 5HT1a receptors have been suggested to couple somewhat selectively to $G\alpha_{i2}$ compared to other $G_{i/o}$ family members.[57] Thus, enhanced $G\alpha_{i2}$ signaling through the $G\alpha_{i2}^{G184S}$ mutation could provide insights in this complex system.

The antidepressant actions of 5HT1a agonists appear to be due to postsynaptic effects.[58,59] One suggested mechanism for the well-known delay in the onset of therapeutic benefits from SSRIs is the desensitization of presynaptic 5HT1a responses which would then permit enhanced 5HT release with a greater postsynaptic antidepressant effect.[59] Since GIRK channels do not mediate presynaptic responses,[56] $G\alpha_{i2}$-mediated GIRK current mechanisms may primarily mediate antidepressant, rather than prodepressant, actions. To test this hypothesis, Talbot *et al.*[28] studied $G\alpha_{i2}^{G184S}$ mutant mice in several preclinical models of antidepressant action. Both heterozygous and homozygous mice showed reduced immobility times in the tail-suspension test as well as decreases in novelty-induced hypophagia—a model that exhibits the requirement for chronic treatment by SSRIs to reverse the deficits. The heterozygotes showed an intermediate phenotype. Immobility times were increased to normal levels by the 5HT1a antagonist WAY-100635 suggesting that the mutants had spontaneous activation of the receptors by endogenous release of serotonin. In the heterozygotes, the actions of both SSRIs and 5HT1a agonists were markedly potentiated compared to control mice.[28] Surprisingly, the 5HT1a receptor-mediated hypothermia response, which has been attributed to presynaptic actions, was not potentiated. This suggests that different physiological processes exhibit markedly different levels of control by RGS proteins. This could provide important selectivity even for a single receptor function upon pharmacological modulation of RGS protein activity.

5. OBSERVED PHENOTYPES WITH $G\alpha_o^{+/G184S}$ KNOCK-IN MICE

There are many overlaps in functional activity of $G\alpha_i$ and $G\alpha_o$ proteins.[4,60] $G\alpha_o$ is highly expressed in the brain and has been strongly implicated in inhibition of presynaptic neurotransmitter release.[6,61] This can be mediated through direct $G\beta\gamma$ inhibition of N- and P/Q-type Ca^{++} channels[6] as well as with the vesicular release machinery.[62–64]

5.1 General Phenotype

$G\alpha_o^{G184S/G184S}$ homozygous mutant mice are nonviable.[22,35] While Mendelian ratios of the three genotypes are observed from heterozygous $G\alpha_o^{+/G184S} \times G\alpha_o^{+/G184S}$ crosses at embryonic day 17.5 (E17.5), there are no viable homozygotes at weaning. Examination at P0 or the day of birth shows a significant number of homozygotes but none survived their first day.[35] The cause is not clear but other early neonatal lethal mutations often cause abnormalities in breathing or feeding. The heterozygous $G\alpha_o^{+/G184S}$ are viable but are modestly under-represented (about 30–50%) compared to the mice carrying the wild-type *Gnao1* gene. There were, however, no gross morphological or behavioral differences between WT and heterozygous mutant RGSi $G\alpha_o$ mice.[22,35]

5.2 Effects on Opioid Signaling

Early studies with the RGSi $G\alpha_o$ mutant in transfected cell systems showed a profound enhancement of mu-opioid signaling.[65] In C6 glioma cells expressing the mu-opioid receptor, there was a nearly 50-fold increase in potency of cAMP inhibition by the full agonist DAMGO while the partial agonist morphine showed a marked enhancement of the maximum suppression of cAMP levels.[65] Lamberts *et al.*[34] demonstrated a substantial increase in both baseline and mu agonist-induced antinociception in the hot-plate assay, which is considered a measure of supraspinal analgesia. Effects in the tail-withdrawal assay were less clear. These results suggest that $G\alpha_o$ does mediate some opioid analgesic actions and that endogenous RGS proteins can suppress them. This conclusion is further supported by use of heterozygous $G\alpha_o^{+/-}$ knock-out mice which showed lower potency of morphine in the hot-plate test.[66]

5.3 *GNAO1* in Epilepsy

There are many studies of the role of G proteins, including $G_{i/o}$, in seizures and epilepsy. Specifically, pertussis toxin treatment has been shown to enhance seizures in the multiple epilepsy models.[67,68] Consistent with this, homozygous $G\alpha_o$ knock-out mutant exhibit seizures.[5] However, the $G\alpha_o$ gain-of-function mutants—G184S—also have a seizure phenotype. The complex regulation of neural excitability, however, makes simple predictions of seizure and epilepsy phenomena very challenging. As described above for chemotaxis, it appears that either decreases or increases in $G\alpha_o$ signaling can result in seizures.

5.4 *GNAO1* G184S Mutants

5.4.1 Slice Studies

The first study of a seizure phenotype using the RGSi $G\alpha_o$ subunit knock-in mice attempted to ask the question, which $G_{i/o}$ protein mediates a known catecholamine-induced suppression of epileptiform activity in hippocampal slices.[33] In this model, picrotoxin was applied to slices to inhibit GABAergic signaling and epileptiform activity in the CA3 region of the hippocampus was measured using extracellular recordings. Epinephrine-mediated activation of the α_{2A} adrenergic receptor strongly suppressed this activity and provided accurate concentration-response measurements. The IC_{50} for epinephrine was 19–23 nM for WT mice. Hippocampal slices from the $G\alpha_{i2}^{+/G184S}$ mutant mice showed an IC_{50} of 19 nM while those from $G\alpha_o^{+/G184S}$ mutant mice gave an IC_{50} of 2.5 nM (i.e., approximately eightfold increased potency vs. WT controls). Thus, the α_{2A} adrenergic receptor-mediated suppression of epileptiform activity in CA3 region of the hippocampus appears to be mediated by $G\alpha_o$ and not $G\alpha_{i2}$. This does not appear to be due to differential expression as all four of the $G_{i/o}$ subtypes show clear expression in the CA3 pyramidal cells in the Allen Brain Atlas.[69] It is possible that this could be due to the role of $G\alpha_o$ in presynaptic inhibition.

5.4.2 In Vivo

In contrast to the results in brain slices data, the $G\alpha_o^{+/G184S}$ mutant mice show a proseizure phenotype. In addition to the perinatal lethality, the $G\alpha_o^{+/G184S}$ mutant mice were found to die prematurely after reaching adulthood. Exploration of this showed that they exhibited seizures before dying and also had a significant increase in spontaneous epileptiform discharges on the EEG.[22,35] They were also sensitized to pentylene tetrazole-induced kindling (i.e., seizures that result from repeated administration of subthreshold excitatory stimulation).

This raises a difficult question. Why do the mice have enhanced antiseizure activity upon epinephrine administration to slices while they exhibit a proseizure behavior *in vivo*? It is important to note that in the slice studies with epinephrine, inhibitory $GABA_A$ signaling was blocked by the picrotoxin. If the $G\alpha_o$ was reducing both excitatory and inhibitory signaling, its effects on the excitatory arm would be magnified by eliminating the inhibitory component. In the absence of the picrotoxin, both inhibitory and excitatory transmitter release would be suppressed by the activation of $G\alpha_o$. If $G\alpha_o$ had a greater effect on inhibitory transmitter release (see model in Fig. 3C), then the balance would shift toward excitatory signaling

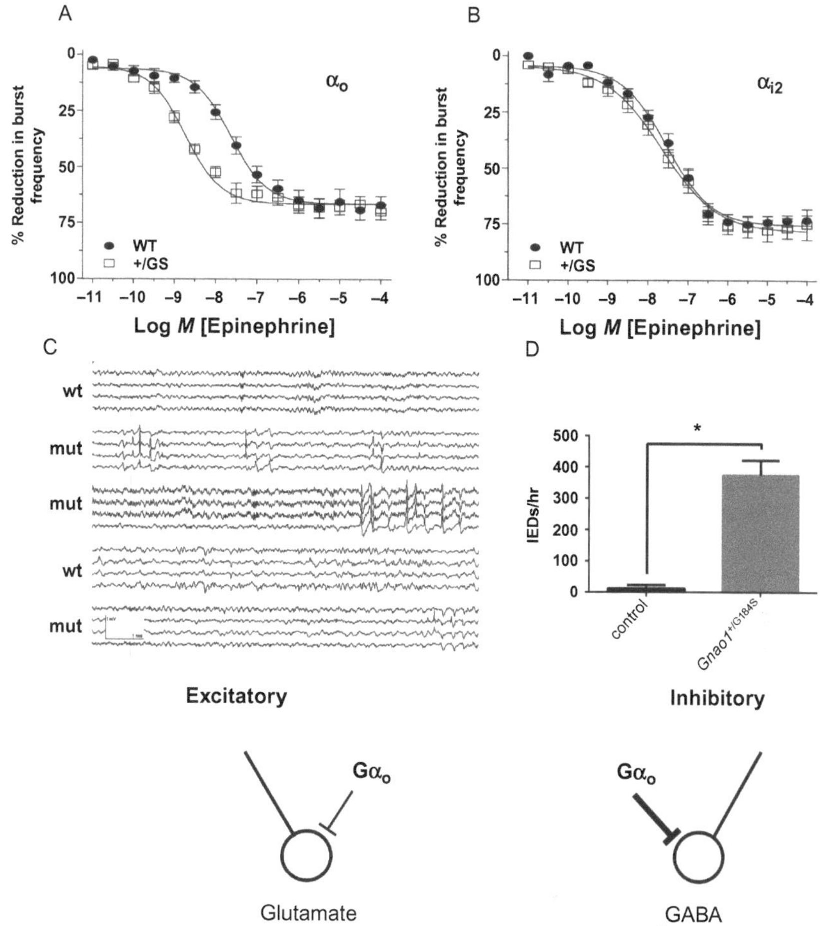

Figure 3 *Gnao1* in epilepsy. Hippocampal slices were obtained from (A) $G\alpha_o^{+/G184S}$ and (B) $G\alpha_{i2}^{+/G184S}$ mutant mice and their wild-type littermates. Extracellular recordings from the CA3 region were made after exposure to picrotoxin to block inhibitory neurotransmission as reported.[33] The frequency of burst activity was measured in the presence of epinephrine, which activates α_{2A} adrenergic receptors to suppress bursting behavior. The RGSi $G\alpha_o$ mutant markedly enhances the potency of epinephrine to suppress bursting behavior in the slices while the $G\alpha_{i2}$ mutant does not. (C and D) $G\alpha_o^{+/G184S}$ mutant mice on the C57BL/6 background were observed to have lethal seizures. EEG recordings of mutant and WT mice showed that mutants had a substantially increased frequency of epileptiform discharges.[35] (E) Model of $G\alpha_o$ in regulation of both excitatory and inhibitory transmitter release. Given the well-known ability of G_o to suppress neurotransmitter release via released $G\beta\gamma$ subunits, a model to rationalize the suppression of epileptiform bursting *in vitro* with an increase *in vivo* is shown. If G_o inhibits both excitatory and inhibitory signaling but has a more pronounced effect on the inhibitory signaling that could explain the results. Specifically, the *in vitro* data in (A) and (B) were measured in the presence of picrotoxin to suppress inhibitory signaling. In that situation, the inhibition of excitation would predominate. When both excitatory and inhibitory signaling are active as is true *in vivo*, then the effect on inhibitory functions would be expected to predominate.

and potentially to the seizure phenotype that was observed.[22,35] These apparently contradictory results underscore the complexity of epilepsy and seizure mechanisms. These results suggest a role for RGS proteins in control of neural excitability and potentially for epilepsy syndromes in humans.

5.5 EIEE17: Human Mutant *GNAO1* Alleles in Epilepsy

In late 2013, a striking report was published by Nakamura *et al.*[70] in which four children with severe early onset epilepsy (OMIM: EIEE17) were found to have heterozygous missense mutations in *GNAO1*. The initial report suggested that these mutations lead to loss of function of the $G\alpha_o$ proteins. Further mutant alleles have been identified in other studies (Epi4K). In the mouse model of $G\alpha_o^{+/G184S}$ mutants, it is most likely that the effect is due to a gain of function. The mechanism of the heterozygous human mutations that produce EIEE17 is not entirely clear. Since they have this effect in the heterozygous context, it could be due to gain-of-function effects, haploinsufficiency, or a dominant negative effect. Further study will be required to assess these possibilities. Heterozygous $G\alpha_o$ knock-out mice do not show a strong seizure phenotype[35] but homozygotes have been reported to exhibit seizures[5], so GOF or dominant negative actions of the EIEE17 mutant *GNAO1* may be the most likely mechanism. It would also be of great interest to know if any RGS mutations could lead to the same phenotype as seen with the $G\alpha_o^{+/G184S}$ mouse mutants.

6. SUMMARY AND CONCLUSIONS

The RGS protein family is large and heterogeneous. Given the overlapping expression and substantial redundancy in function of the 20 RGS proteins, it is not surprising that most individual RGS knockouts have modest phenotypes. The RGSi $G\alpha$ subunit mutants described here provide an alternative approach to assessing the roles of RGS proteins in physiological and cellular functions. They also have provided a novel tool to dissect the roles of the different members of the $G_{i/o}$ family. Some of the observed phenotypes are potentially beneficial including, for $G\alpha_{i2}$, an antidepressant-like effect, reduced weight gain and enhanced insulin sensitivity, and protection against cardiac reperfusion injury. In contrast, other effects are detrimental including enhanced cardiac fibrosis for $G\alpha_{i2}$ and an epileptic phenotype for $G\alpha_o$. These mouse models can provide insights into potential therapeutic strategies via targeting RGS proteins with chemical inhibitors[2,17,19,71] but they also may reveal novel pathologic mechanisms as in the epilepsy model with $G\alpha_o$. Further study of these mutant models, especially with other Gα subunit subtypes and/or conditional mutant systems should be quite revealing.

REFERENCES

1. Wettschureck N, Offermanns S. Mammalian G proteins and their cell type specific functions. *Physiol Rev*. 2005;85(4):1159–1204.
2. Neubig RR, Siderovski DP. Regulators of G-protein signalling as new central nervous system drug targets. *Nat Rev Drug Discov*. 2002;1(3):187–197.
3. Berman DM, Wilkie TM, Gilman AG. GAIP and RGS4 are GTPase-activating proteins for the Gi subfamily of G protein alpha subunits. *Cell*. 1996;86(3):445–452.
4. Sunahara RK, Taussig R. Isoforms of mammalian adenylyl cyclase: multiplicities of signaling. *Mol Interv*. 2002;2(3):168–184.
5. Valenzuela D, Han X, Mende U, et al. G alpha(o) is necessary for muscarinic regulation of Ca^{2+} channels in mouse heart. *Proc Natl Acad Sci U S A*. 1997;94(5):1727–1732.
6. Tedford HW, Zamponi GW. Direct G protein modulation of Cav2 calcium channels. *Pharmacol Rev*. 2006;58(4):837–862.
7. Chan RK, Otte CA. Isolation and genetic analysis of *Saccharomyces cerevisiae* mutants supersensitive to G1 arrest by a factor and alpha factor pheromones. *Mol Cell Biol*. 1982;2(1):11–20.
8. Dohlman HG, Song J, Ma D, Courchesne WE, Thorner J. Sst2, a negative regulator of pheromone signaling in the yeast *Saccharomyces cerevisiae*: expression, localization, and genetic interaction and physical association with Gpa1 (the G-protein alpha subunit). *Mol Cell Biol*. 1996;16(9):5194–5209.
9. Koelle MR, Horvitz HR. EGL-10 regulates G protein signaling in the *C. elegans* nervous system and shares a conserved domain with many mammalian proteins. *Cell*. 1996;84(1):115–125.
10. Druey KM, Blumer KJ, Kang VH, Kehrl JH. Inhibition of G-protein-mediated MAP kinase activation by a new mammalian gene family. *Nature*. 1996;379(6567): 742–746.
11. Chen CK, Burns ME, He W, Wensel TG, Baylor DA, Simon MI. Slowed recovery of rod photoresponse in mice lacking the GTPase accelerating protein RGS9-1. *Nature*. 2000;403(6769):557–560.
12. Heximer SP, Knutsen RH, Sun X, et al. Hypertension and prolonged vasoconstrictor signaling in RGS2-deficient mice. *J Clin Invest*. 2003;111(4):445–452.
13. Rahman Z, Schwarz J, Gold SJ, et al. RGS9 modulates dopamine signaling in the basal ganglia. *Neuron*. 2003;38(6):941–952.
14. Kaur K, Kehrl JM, Charbeneau RA, Neubig RR. RGS-insensitive Galpha subunits: probes of Galpha subtype-selective signaling and physiological functions of RGS proteins. *Methods Mol Biol*. 2011;756:75–98.
15. Nishiguchi KM, Sandberg MA, Kooijman AC, et al. Defects in RGS9 or its anchor protein R9AP in patients with slow photoreceptor deactivation. *Nature*. 2004;427(6969): 75–78.
16. Yang J, Kamide K, Kokubo Y, et al. Genetic variations of regulator of G-protein signaling 2 in hypertensive patients and in the general population. *J Hypertens*. 2005; 23(8):1497–1505.

17. Blazer LL, Neubig RR. Small molecule protein-protein interaction inhibitors as CNS therapeutic agents: current progress and future hurdles. *Neuropsychopharmacology.* 2009;34(1):126–141.
18. Sjogren B, Neubig RR. Thinking outside of the (RGS) box: new approaches to therapeutic targeting of regulators of G protein signaling. *Mol Pharmacol.* 2010;78(4): 550–557.
19. Zhong H, Neubig RR. Regulator of G protein signaling proteins: novel multifunctional drug targets. *J Pharmacol Exp Ther.* 2001;297(3):837–845.
20. DiBello PR, Garrison TR, Apanovitch DM, et al. Selective uncoupling of RGS action by a single point mutation in the G protein alpha-subunit. *J Biol Chem.* 1998;273(10): 5780–5784.
21. Lan KL, Sarvazyan NA, Taussig R, et al. A point mutation in Galphao and Galphai1 blocks interaction with regulator of G protein signaling proteins. *J Biol Chem.* 1998;273(21):12794–12797.
22. Kehrl JH. *Pharmacology.* Ann Arbor, MI: University of Michigan; 2014.
23. Huang X, Fu Y, Charbeneau RA, et al. Pleiotropic phenotype of a genomic knock-in of an RGS-insensitive G184S Gnai2 allele. *Mol Cell Biol.* 2006;26(18):6870–6879.
24. Fu Y, Huang X, Piao L, Lopatin AN, Neubig RR. Endogenous RGS proteins modulate SA and AV nodal functions in isolated heart: implications for sick sinus syndrome and AV block. *Am J Physiol Heart Circ Physiol.* 2007;292(5):H2532–H2539.
25. Huang X, Charbeneau RA, Fu Y, et al. Resistance to diet-induced obesity and improved insulin sensitivity in mice with a regulator of G protein signaling-insensitive G184S Gnai2 allele. *Diabetes.* 2008;57(1):77–85.
26. Icaza EE, Huang X, Fu Y, Neubig RR, Baghdoyan HA, Lydic R. Isoflurane-induced changes in righting response and breathing are modulated by RGS proteins. *Anesth Analg* 2009;109(5):1500–1505.
27. Signarvic RS, Cierniewska A, Stalker TJ, et al. RGS/Gi2alpha interactions modulate platelet accumulation and thrombus formation at sites of vascular injury. *Blood.* 2010;116(26):6092–6100.
28. Talbot JN, Jutkiewicz EM, Graves SM, et al. RGS inhibition at G(alpha)i2 selectively potentiates 5-HT1A-mediated antidepressant effects. *Proc Natl Acad Sci U S A.* 2010;107(24):11086–11091.
29. Waterson RE, Thompson CG, Mabe NW, et al. Galpha(i2)-mediated protection from ischaemic injury is modulated by endogenous RGS proteins in the mouse heart. *Cardiovasc Res.* 2011;91(1):45–52.
30. Kaur K, Parra S, Chen R, et al. Galphai2 signaling: friend or foe in cardiac injury and heart failure? *Naunyn Schmiedebergs Arch Pharmacol.* 2012;385(5):443–453.
31. Li P, Neubig RR, Zingarelli B, et al. Toll-like receptor-induced inflammatory cytokines are suppressed by gain of function or overexpression of Galpha(i2) protein. *Inflammation.* 2012;35(5):1611–1617.
32. Parra S, Huang X, Charbeneau RA, et al. Conditional disruption of interactions between Galphai2 and regulator of G protein signaling (RGS) proteins protects the heart from ischemic injury. *BMC Pharmacol Toxicol.* 2014;15:29.
33. Goldenstein BL, Nelson BW, Xu K, et al. Regulator of G protein signaling protein suppression of Galphao protein-mediated alpha2A adrenergic receptor inhibition of mouse hippocampal CA3 epileptiform activity. *Mol Pharmacol.* 2009;75(5):1222–1230.
34. Lamberts JT, Smith CE, Li MH, Ingram SL, Neubig RR, Traynor JR. Differential control of opioid antinociception to thermal stimuli in a knock-in mouse expressing regulator of G-protein signaling-insensitive Galphao protein. *J Neurosci.* 2013;33(10): 4369–4377.
35. Kehrl JM, Sahaya K, Dalton HM, et al. Gain-of-function mutation in *Gnao1*: a murine model of epileptiform encephalopathy (EIEE17)? *Mamm Genome.* 2014;25:202–210.

36. Neer EJ, Clapham DE. Signal transduction through G proteins in the cardiac myocyte. *Trends Cardiovasc Med.* 1992;2(1):6–11.
37. Logothetis DE, Kurachi Y, Galper J, Neer EJ, Clapham DE. The beta gamma subunits of GTP-binding proteins activate the muscarinic K^+ channel in heart. *Nature.* 1987;325(6102):321–326.
38. Wickman K, Nemec J, Gendler SJ, Clapham DE. Abnormal heart rate regulation in GIRK4 knockout mice. *Neuron.* 1998;20(1):103–114.
39. Sowell MO, Ye C, Ricupero DA, et al. Targeted inactivation of alphai2 or alphai3 disrupts activation of the cardiac muscarinic K+ channel, IK+Ach, in intact cells. *Proc Natl Acad Sci U S A.* 1997;94(15):7921–7926.
40. Doupnik CA, Davidson N, Lester HA, Kofuji P. RGS proteins reconstitute the rapid gating kinetics of gbetagamma-activated inwardly rectifying K+ channels. *Proc Natl Acad Sci U S A.* 1997;94(19):10461–10466.
41. Fu Y, Huang X, Zhong H, Mortensen RM, D'Alecy LG, Neubig RR. Endogenous RGS proteins and Galpha subtypes differentially control muscarinic and adenosine-mediated chronotropic effects. *Circ Res.* 2006;98(5):659–666.
42. Waterson RE, Thompson CG, Mabe NW, et al. Galphai2-mediated protection from ischaemic injury is modulated by endogenous RGS proteins in the mouse heart. *Cardiovasc Res.* 2011;91(1):45–52.
43. Foerster K, Groner F, Matthes J, Koch WJ, Birnbaumer L, Herzig S. Cardioprotection specific for the G protein Gi2 in chronic adrenergic signaling through beta 2-adrenoceptors. *Proc Natl Acad Sci U S A.* 2003;100(24):14475–14480.
44. Chesley A, Lundberg MS, Asai T, et al. The beta(2)-adrenergic receptor delivers an antiapoptotic signal to cardiac myocytes through G(i)-dependent coupling to phosphatidylinositol 3′-kinase. *Circ Res.* 2000;87(12):1172–1179.
45. Weber KT, Jalil JE, Janicki JS, Pick R. Myocardial collagen remodeling in pressure overload hypertrophy. A case for interstitial heart disease. *Am J Hypertens.* 1989; 2(12 Pt 1):931–940.
46. Wiege K, Le DD, Syed SN, et al. Defective macrophage migration in Galphai2- but not Galphai3-deficient mice. *J Immunol.* 2012;189(2):980–987.
47. Zarbock A, Deem TL, Burcin TL, Ley K. Galphai2 is required for chemokine-induced neutrophil arrest. *Blood.* 2007;110(10):3773–3779.
48. Hwang IY, Park C, Kehrl JH. Impaired trafficking of Gnai2+/− and Gnai2−/− T lymphocytes: implications for T cell movement within lymph nodes. *J Immunol.* 2007;179(1):439–448.
49. Dalwadi H, Wei B, Schrage M, et al. B cell developmental requirement for the G alpha i2 gene. *J Immunol.* 2003;170(4):1707–1715.
50. Hwang IY, Park C, Harrison K, Boularan C, Gales C, Kehrl JH. An essential role for RGS protein/Galphai2 interactions in B lymphocyte-directed cell migration and trafficking. *J Immunol.* 2015;194(5):2128–2139.
51. Cho H, Kamenyeva O, Yung S, et al. The loss of RGS protein-Galpha(i2) interactions results in markedly impaired mouse neutrophil trafficking to inflammatory sites. *Mol Cell Biol.* 2012;32(22):4561–4571.
52. Hwang IY, Park C, Harrision KA, Huang NN, Kehrl JH. Variations in Gnai2 and Rgs1 expression affect chemokine receptor signaling and the organization of secondary lymphoid organs. *Genes Immun.* 2010;11(5):384–396.
53. Han SB, Moratz C, Huang NN, et al. Rgs1 and Gnai2 regulate the entrance of B lymphocytes into lymph nodes and B cell motility within lymph node follicles. *Immunity.* 2005;22(3):343–354.
54. Huang TT, Zong Y, Dalwadi H, et al. TCR-mediated hyper-responsiveness of autoimmune Galphai2(−/−) mice is an intrinsic naive CD4(+) T cell disorder selective for the Galphai2 subunit. *Int Immunol.* 2003;15(11):1359–1367.

55. Hornquist CE, Lu X, Rogers-Fani PM, et al. G(alpha)i2-deficient mice with colitis exhibit a local increase in memory CD4+ T cells and proinflammatory Th1-type cytokines. *J Immunol.* 1997;158(3):1068–1077.
56. Luscher C, Jan LY, Stoffel M, Malenka RC, Nicoll RA. G protein-coupled inwardly rectifying K+ channels (GIRKs) mediate postsynaptic but not presynaptic transmitter actions in hippocampal neurons. *Neuron.* 1997;19(3):687–695.
57. Raymond JR, Mukhin YV, Gelasco A, et al. Multiplicity of mechanisms of serotonin receptor signal transduction. *Pharmacol Ther.* 2001;92(2–3):179–212.
58. Martin P, Beninger RJ, Hamon M, Puech AJ. Antidepressant-like action of 8-OH-DPAT, a 5-HT1A agonist, in the learned helplessness paradigm: evidence for a postsynaptic mechanism. *Behav Brain Res.* 1990;38(2):135–144.
59. Celada P, Bortolozzi A, Artigas F. Serotonin 5-HT1A receptors as targets for agents to treat psychiatric disorders: rationale and current status of research. *CNS Drugs.* 2013;27(9):703–716.
60. Mark MD, Herlitze S. G-protein mediated gating of inward-rectifier K+ channels. *Eur J Biochem.* 2000;267(19):5830–5836.
61. Chen H, Lambert NA. Endogenous regulators of G protein signaling proteins regulate presynaptic inhibition at rat hippocampal synapses. *Proc Natl Acad Sci U S A.* 2000;97(23): 12810–12815.
62. Wells CA, Zurawski Z, Betke KM, et al. Gbetagamma inhibits exocytosis via interaction with critical residues on soluble N-ethylmaleimide-sensitive factor attachment protein-25. *Mol Pharmacol.* 2012;82(6):1136–1149.
63. Gerachshenko T, Blackmer T, Yoon EJ, Bartleson C, Hamm HE, Alford S. Gbetagamma acts at the C terminus of SNAP-25 to mediate presynaptic inhibition. *Nat Neurosci.* 2005;8(5):597–605.
64. Blackmer T, Larsen EC, Takahashi M, Martin TF, Alford S, Hamm HE. G protein betagamma subunit-mediated presynaptic inhibition: regulation of exocytotic fusion downstream of Ca^{2+} entry. *Science.* 2001;292(5515):293–297.
65. Clark MJ, Harrison C, Zhong H, Neubig RR, Traynor JR. Endogenous RGS protein action modulates mu-opioid signaling through Galphao. Effects on adenylyl cyclase, extracellular signal-regulated kinases, and intracellular calcium pathways. *J Biol Chem.* 2003;278(11):9418–9425.
66. Lamberts JT, Jutkiewicz EM, Mortensen RM, Traynor JR. Mu-opioid receptor coupling to Galpha(o) plays an important role in opioid antinociception. *Neuropsychopharmacology.* 2011;36(10):2041–2053.
67. Mazarati A, Lu X. Regulation of limbic status epilepticus by hippocampal galanin type 1 and type 2 receptors. *Neuropeptides.* 2005;39(3):277–280.
68. Ormandy GC, Jope RS. Pertussis toxin potentiates seizures induced by pilocarpine, kainic acid and N-methyl-D-aspartate. *Brain Res.* 1991;553(1):51–57.
69. Lein ES, Hawrylycz MJ, Ao N, et al. Genome-wide atlas of gene expression in the adult mouse brain. *Nature.* 2007;445(7124):168–176.
70. Nakamura K, Kodera H, Akita T, et al. De novo mutations in *GNAO1*, encoding a Galphao subunit of heterotrimeric G proteins, cause epileptic encephalopathy. *Am J Hum Genet.* 2013;93(3):496–505.
71. Sjogren B, Blazer LL, Neubig RR. Regulators of G protein signaling proteins as targets for drug discovery. *Prog Mol Biol Transl Sci.* 2010;91:81–119.

CHAPTER THREE

RGS Protein Regulation of Phototransduction

Ching-Kang Jason Chen[1,2]
Department of Ophthalmology, Baylor College of Medicine, Houston, Texas, USA
Department of Biochemistry and Molecular Biology, Baylor College of Medicine, Houston, Texas, USA
Department of Neuroscience, Baylor College of Medicine, Houston, Texas, USA
[2]Corresponding author: e-mail address: Ching-Kang.Chen@bcm.edu

Contents

Abstract

First identified in yeast and worm and later in other species, the physiological importance of Regulators of G-protein Signaling (RGS) in mammals was first demonstrated at the turn of the century in mouse retinal photoreceptors, in which RGS9 is needed for timely recovery of rod phototransduction. The role of RGS in vision has also been established a synapse away in retinal depolarizing bipolar cells (DBCs), where RGS7 and RGS11 work redundantly and in a complex with Gβ5-S as GAPs for Goα in the metabotropic glutamate receptor 6 pathway situated at DBC dendritic tips. Much less is known on how RGS protein subserves vision in the rest of the visual system. The research into the roles of RGS proteins in vision holds great potential for many exciting new discoveries.

1. INTRODUCTION

Being pushed out of the brain and into the eye during development, retina lines the back of the eye and is an extension of the central nervous

[1] Ching-Kang Jason Chen is the Alice R. McPherson Retina Research Foundation Endowed Chair at Baylor College of Medicine.

Progress in Molecular Biology and Translational Science, Volume 133
ISSN 1877-1173
http://dx.doi.org/10.1016/bs.pmbts.2015.02.004

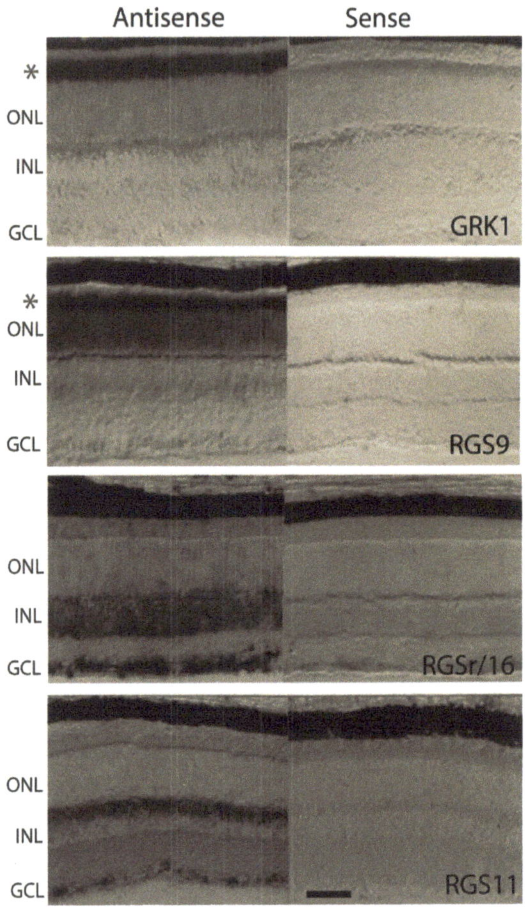

Figure 1 *In situ* hybridization showing localization of GRK1, RGS9, RGSr/16, and RGS11 messages in mouse retinal cross sections. The three nuclear layers: outer nuclear layer (ONL), inner nuclear layer (INL), and ganglion cell layer (GCL) are marked at left. GRK1 message is abundantly present at the inner segment (IS) layer of the retina (red asterisk) and is used as a marker for any potential photoreceptor-specific genes. Sense probes control for degree of background staining. For RGS genes tested for candidacy as a transducin GAP, only RGS9 message appears in the IS and ONL. RGSr/16 is present throughout INL and GCL but notably absent in IS or ONL. RGS11 is localized to some GCL cells and the upper-tier cells in INL. Scale bar equals 50 μm. (See the color plate.)

system. Mature retina has a beautifully layered laminated structure with three nuclear layers (Fig. 1) and subserves vision by converting light into electrical signals in photoreceptors and by processing and encoding changes in light intensities and wavelengths in ambient environment in the rest of retinal neurons. Because of the ease of isolation and maintenance in culture, retina has been extensively studied but as with any mature field, the more we know, the more areas we know we do not know. Since the late 1980s, knowledge on the development and functions of the retina has benefited from the identification of numerous causative mutations in patients with hereditary blinding diseases. Subsequent recapitulation of pathologic conditions of some diseases in genetically engineered model organisms enables trials to use the knowledge gained to assist the care and/or treatment of some patients. With regard to the Regulators of G-protein Signaling (RGS), retinal photoreceptor is where the importance of this gene family in mammalian physiology first demonstrated. This chapter reflects on how the phototransduction field discovered the importance of RGS proteins and describes current state of knowledge about this gene family in vision.

2. FROM PHOTON TO A NEURAL SIGNAL: THE WONDER OF PHOTOTRANSDUCTION

In the heyday of apprenticeship in classical biochemistry, the author's typical day would start with driving to a local slaughterhouse to purchase fresh bovine eyes, usually tens upward to hundreds, keeping them in the dark and on ice, driving back to the lab, dissecting retinas out in a darkroom, and spending the rest of the day running centrifugation rounds to isolate a structure in the retina high in lipid content called the outer segment.[1] With tubes of outer segment preparation in hand, the party began by adding various reagents in the dark and then exposing them to light to see what might happen. Very often we obtained outer segments from hundreds of bovine eyes and fractionated the proteins by column chromatography or other means to purify the protein(s) with the desired activity. More often than not, experimental conditions would be altered to see if and how the activity was enhanced or killed. As one would have guessed, there are many interesting light-dependent reactions in those tubes! The most popular one in the lab at that time was light-dependent activation of phosphodiesterase,[2] assayable by hydrolysis of exogenous radioactive cGMP.[3] We did it frequently because at that time the phototransduction field has come out of the shadow of the calcium hypothesis[4] and firmly established that cGMP is the second messenger used by photoreceptors to transduce light into an electrical signal.[5] The enzymatic cascade found in the outer segment preparations which links photon absorption to the hydrolysis of cGMP is called phototransduction and is a canonical heterotrimeric G-protein signaling pathway (Fig. 2 and for a more detailed review, see Ref. 6). This cascade

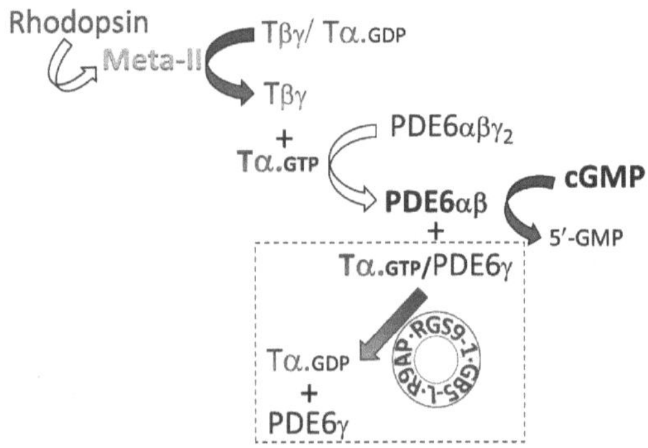

Figure 2 Vertebrate rod phototransduction cascade. Photon absorption by rhodopsin leads to the formation of metarhodopsin II (Meta-II), which catalyzes guanine nucleotide exchange on transducin α subunit (Tα) and results in its dissociation from the βγ dimer (Tβγ). The GTP-bound Tα (Tα.GTP) interacts with PDE6γ to relieve its inhibition on PDE6αβ, freeing this near-perfect phosphodiesterase to hydrolyze cGMP into 5′-GMP. Two gain steps marked by filled ribbons endow rods with the sensitivity to detect single photons. This chapter deals with one reaction that must occur during recovery of rod phototransduction as highlighted in the dashed box. The reaction is the hydrolysis of bound GTP by Tα to GDP, which is assisted *in vivo* by a GTPase-accelerating protein (GAP), now known to be a protein complex consists of RGS9-1, Gβ5-L, and R9AP. Without the GAP complex, Tα turns over bound GTP in seconds, a pace too slow to sustain normal vision. (See the color plate.)

of events starts when rhodopsin absorbs a photon and the chromophore 11-*cis*-retinaldehyde (11-*cis*-retinal) is photoisomerized into the all-*trans* form. Rhodopsin is a G protein-coupled receptor and the covalently linked 11-*cis*-retinal is an inverse agonist. When 11-*cis*-retinal turns into all-*trans* retinal, rhodopsin undergoes a series of structural changes to adopt an active intermediary conformation called Meta-II (R*). R* can catalyze the exchange of GTP for GDP on the alpha subunit of the photoreceptor-specific G protein called transducin (Tα). During its short active lifetime, R* activates tens of transducins and hence provides the initial gain to the cascade. When Tα is bound to GDP, it associates tightly with the βγ subunits (Tβγ) to form a heterotrimer. The binding of GTP causes Tα to dissociate from Tβγ and subsequently binds the inhibitory subunit of the cGMP-specific phosphodiesterase (PDE6γ). The sequestration of PDE6γ by Tα relieves inhibition on the catalytic αβ subunit (PDE6αβ). Uninhibited PDE6αβ is a near-perfect enzyme, with the rate of catalysis limited only by diffusion of substrate into the catalytic site. The catalytic activity of PDE6 constitutes the major gain step in phototransduction, which results in a rapid decline of cGMP level inside the outer segment. In the dark, the cGMP-gated (CNG) nonselective cation channels on the plasma membrane stay open by binding to cGMP. The drop in cGMP level as a result of phototransduction closes many CNG channels and membrane potential decreases from −40 to about −70 mV. This light-induced hyperpolarization spreads from the outer segment to the rest of the cell, alters voltage-dependent channel activity along the way, and eventually blocks the release of glutamate at photoreceptor synaptic terminal.[7] This amazing chain of events takes place within a few hundred milliseconds after photoactivation of rhodopsin and can be captured in real time in several recording configurations using pulled glass microelectrodes.[8–10] Macroscopically, because of the layered retinal structure, phototransduction can also be picked up using a noninvasive transcorneal field potential recording technique called electroretinography (ERG), where it appears under bright stimulus conditions as the so-called A-wave. The biochemical scheme of phototransduction is simple but effective in empowering a rod cell to detect single photons, while allowing a cone cell to transduce light with faster kinetics but lesser sensitivity. Using rods and cones and connecting them to a network of not yet very well understood endogenous neuronal circuits, the retina operates with an amazing dynamic range of ten orders of magnitudes in stimulus intensity, a task unmatched in any contemporary manmade devices.

3. THE NEED FOR SPEED: DISCREPANCY ON G-PROTEIN SHUTOFF DURING PHOTOTRANSDUCTION RECOVERY

In early 1990s, the activation phase of vertebrate phototransduction was considered solved because differences in sensitivity and kinetics in rods of various vertebrate species could be nicely modeled when body temperatures and photoreceptor sizes were taken into consideration.[11] The field started to tackle a more difficult question: how phototransduction is turned off.[12] It was considered difficult because unlike the sequential activation of known enzymes during activation, to turn phototransduction off in a timely manner in just a few hundred milliseconds, all active intermediates such as R*, GTP-bound Tα, and PDE6αβ accumulated during activation need to be deactivated and the cGMP level must be restored to reopen CNG channels. The utility of several assayable light-dependent reactions in outer segment preparations was realized in this context, such as light-dependent phosphorylation of rhodopsin[13] and change in membrane affinity of a protein called arrestin.[14] Fast forward two decades and thanks to all assiduous vision researchers, now we have a clearer picture about what is going on during phototransduction recovery. The R* got phosphorylated at its C-terminal tail Ser and Thr residues[15] by an enzyme called rhodopsin kinase (aka G protein-coupled receptor kinase 1 (GRK1))[16] and arrestin binds the phosphorylated R* to prevent it from further interaction with transducin.[17] As a G protein, Tα has an intrinsic GTPase activity that hydrolyzes bound GTP to GDP and when it occurs, GDP-bound Tα dissociates from PDE6γ and reassociates with Tβγ subunits. Freed PDE6γ reinhibits PDE6αβ to its basal activity, allowing reaccumulation of cGMP inside the outer segment. To replenish cGMP in a timely manner to open CNG channels, membrane-bound guanylate cyclase (GC) activity is activated to synthesize cGMP *de novo* from GTP.[18] The timing of heightened cGMP synthesis is governed by a light-dependent decline of calcium concentration inside the outer segment[19] and is mediated by a small calcium-binding protein called GCAP (guanylate cyclase-activating protein)[20,21] that in its calcium bound form binds and inhibits the membrane-bound GCs. As mentioned above, all reactions used during phototransduction recovery are assayable and hence kinetics of individual reactions can be measured under chemically defined conditions using purified enzymes. It is under the strict mandate of classical biochemistry that a glaring discrepancy concerning transducin deactivation surfaced. This is because it takes seconds for purified Tα to hydrolyze bound GTP to GDP in the test tube, but recovery of phototransduction is finished within a few hundred milliseconds[22]! This suggests that GTP hydrolysis by Tα is somehow accelerated in the outer segment. Given the availability of material and the ease of assaying GTP hydrolysis, there was a fury in the field to find the putative GAP (GTPase-accelerating protein). However, the traditional biochemical approach used so successfully in the field quickly revealed some bad news given that the GAP activity is membrane associated and became labile in the presence of detergents, making it formidable to study it using conventional methods. However, these efforts did produce some leads in that PDE6γ can enhance the GAP activity even though it is by itself not the GAP.[23,24] Around that time, a new group of proteins coined Regulators of G-protein Signaling (RGS) were identified by forward genetic screens in yeast and worm, and later found to be abundantly present in mammals.[25,26] These RGS proteins are simple negative regulators of heterotrimeric G-proteins in yeast and worm, but in mammals they exist with greater diversity in sizes and expression patterns in different tissues.[27] All RGS proteins have a conserved RGS domain of approximately 120 amino acids in length, which is necessary and sufficient for their GAP activity toward Gi/o proteins.[28] This new insight set off another fury in the phototransduction field to test whether the long sought-after transducin GAP might after all be an RGS protein. Several labs used degenerate oligonucleotides to screen retinal cDNA library for RGS transcripts and found a great deal of diversity in the retina. Several members of this family, such as RGSr/16, GAIP (G alpha interacting protein), RGS9, RGS4, and Ret-RGS1, were further tested for their GAP activity toward purified transducin *in vitro* and surprisingly they all possessed GAP activity, albeit to varying degrees.[29–34] Could it be possible that multiple RGS proteins, instead of a pivotal one, are present in the outer segment to ensure timely shutoff of transducin? To gain further insight, additional criteria such as membrane affinity, photoreceptor-specific expression pattern, and whether GAP activity could be enhanced *in vitro* by exogenous PDE6γ were considered. Among them, the telltale sign for some was whether any of these RGS proteins are similarly expressed in a photoreceptor-specific manner. Most, if not all, proteins involved in phototransduction such as rhodopsin, transducin, and rhodopsin kinase are exclusively expressed in photoreceptors. Taken all into account, RGS9 stood out as a promising candidate because of its photoreceptor-specific expression pattern (Fig. 1), while other candidates such as RGSr/16 and RGS11 were located elsewhere in the retina (Fig. 1).[35,36] Furthermore, RGSr/16's GAP activity toward transducin was inhibited rather than enhanced by PDE6γ,[37] making it highly unlikely that it is a physiological GAP for transducin. To test whether RGS9 is indeed the GAP, rather than one of the GAPs for transducin in photoreceptors, Chen *et al.* inactivated it and found that recovery in rod and cone was severely delayed in RGS9 knockout mice.[38,39] A few years later, Nishiguchi *et al.* reported sporadic human cases of a novel ophthalmic condition called bradyopsia, where recessive mutations in RGS9 or its membrane anchoring protein (R9AP, see below) render the patients with difficulty adapting to sudden changes in luminance levels and unable to see fast-moving objects in low-contrast conditions.[40] Since then, more patients with similar conditions caused by loss-of-function mutations in RGS9 were found.[41–43] A two-decade worth of earnest efforts to solve a long-standing controversy in phototransduction recovery brings to light the importance of RGS proteins in human biology and disease etiology. This is one of the reasons why hypothesis-driven basic research aiming to solve a mystery is always a good bet for funding agencies, because if not supported in a timely manner, many opportunities for exciting new discoveries would have been missed.

4. CANNOT DO IT ALONE: THE TRANSDUCIN GAP IS A PROTEIN COMPLEX

Strictly speaking, RGS9 is not a photoreceptor-specific gene because it is also transcribed in other CNS regions including striatum,[44,45] although alternative splicing there generates a larger protein, RGS9-2, with a different and bigger C-terminal domain, as opposed to the photoreceptor-specific splice form RGS9-1 with a shorter "retina-specific" C-terminal tail.[46] The gene targeting approach in laboratory mice used to settle whether transducin is "gapped" by one or many RGS proteins has around that time identified many genes critical for phototransduction.[16,17,47–49] This reverse genetic approach sometime produces unexpected results. In the case of transducin GAP, one surprise is that RGS9-1 has two *in vivo* partners, namely Gβ5-L[50] and R9AP (RGS9 anchoring protein).[51] They associate tightly with RGS9-1 in photoreceptors to form a ternary protein complex. Gβ5-L binds and guides RGS9-1 to interact with R9AP, while the transmembrane domain of R9AP ensures the entire complex is membrane associated. In mice lacking RGS9, Gβ5-L becomes unstable and likewise, in mice lacking Gβ5, RGS9-1 becomes unstable. These studies revealed a unique obligate partnership between RGS9-1 and Gβ5-L, suggesting first during biochemical characterization of then elusive transducin GAP,[52] wherein the loss of one leads to the instability of the other despite the presence of messenger RNA. Not surprisingly, the recovery delays observed in rods without RGS9 are also apparent in rods lacking Gβ5 or R9AP.[53,54] Equally if not more interesting is the higher expression level of this GAP complex in cone than in rod.[55] Depending on species, this difference can be as high as 10-fold.[45,56] As mentioned earlier, rod is more sensitive to cone but responds to light in a slower fashion. The higher level of transducin GAP in cone has been suggested to be a deciding factor governing kinetic differences between these two photoreceptors. To test the idea, Chen *et al.* attempted to transgenically overexpress RGS9-1 in rod using a construct putting RGS9-1 cDNA downstream of a rhodopsin promoter. This effort failed.[57] Similar attempts by overexpressing the obligate partner Gβ5-L, by itself or in combination with overexpressing RGS9-1, were also unsuccessful in raising photoreceptor GAP level. This remained a puzzle for a few years until the anchoring protein R9AP was inactivated; it precipitously became clear that overall GAP expression level is determined in rods by R9AP expression alone as in R9AP heterozygous knockout animals transducin GAP level is halved,[53] while corresponding heterozygous knockouts in either RGS9 or Gβ5 have normal GAP level.[38,57] Krispel *et al.* again put this notion to test by overexpressing R9AP using a rhodopsin promoter and found indeed that R9AP overexpression can significantly elevate transducin GAP level in rods.[58] What was even more interesting, as the focus of the investigation turned, was that photoresponses in rods with overexpressed GAP recover much faster than their wild-type controls and do so in a dose-dependent manner to GAP levels. Because raising GAP level altered rod photoresponse waveform with a faster rate of return to baseline, these findings indicate that GAP level is indeed a determining factor in kinetic differences between rod and cone.[59] This "gain-of-function" type of experiment is the first successful case in the vision field to confer a better response property to a specific cell type and thus has a wide applicability to other retinal cell types in the quest to understand their contributions and functions in the retina. It is now possible to ask whether response duration of rod rate-limits temporal resolution in scotopic vision under normal physiological conditions. The then powerful and now routine "loss-of-function" approach like gene knockouts merely generates an artificial limiting step and hence has little or no power to shed light on what is normally limiting. In this regard, while RGS9-1/Gβ5-L/R9AP function as the transducin GAP is firmly established, what other RGS proteins may be doing in vision becomes very intriguing. Examining the roles of RGS in vision is thus a very fertile ground with the potential for many exciting new discoveries.

5. TRANSLOCATION AND REGULATION OF RGS9-1

Several proteins involved in phototransduction translocate in and out of the outer segment compartment in a light-dependent manner (for a review, see Ref. 60). Translocation of transducin and arrestin was first noted in the early 1980s and then confirmed beyond the caveat of epitope masking shortly after the turn of the century.[61] The speculated roles of protein movement are in protecting photoreceptors from light-dependent degeneration and/or long-term light adaptation. A photoreceptor enriched calcium-binding protein, recoverin, which binds and inhibits rhodopsin kinase,[62,63] also exhibits noticeable light-dependent movement out of the outer segment,[62] although its physiological function is less clear. Given that rod photoresponse duration is controlled by transducin GAP level, it is of interest to note that RGS9-1 was recently reported to reside mainly in the inner segment under prolonged dark adaptation and move swiftly out to the outer segment upon illumination.[64] The reported translocation of RGS9-1 from inner to outer segment occurred at much dimmer illumination than those required to translocate transducin or arrestin. Interestingly, of the three components of the transducin GAP, only RGS9-1 and Gβ5-L exhibit this peculiar translocation phenomenon and the anchoring protein R9AP stays in the outer segment regardless of illumination level. While phenomenological, it was also found that under prolonged dark adaptation RGS9-1 is phosphorylated at S475[65] (presumably by PKCs[66]). Light exposure leads to a rapid decrease in S475 phosphorylation, promoting the assembly of the ternary protein complex and translocation to outer segment.[64] This finding is somewhat at odd with the well-documented stereotypic nature of rod single-photon responses.[67] Should translocation of RGS9-1 and Gβ5-L in fact occur, one would expect the single-photon responses to have some degree of variability in recovery phase. However, should RGS9-1 translocation occur quickly, as shown in just a few minutes upon dim light exposure,[64] only the initial few recorded light responses will deviate from the ensemble responses typically collected hundreds of times every 5–10 s for a period of 30+ min during suction pipette recordings. Clearly, there is much work to be done here to solve this controversy. Finally, with regard to the phosphorylation of RGS9-1, there appears to be other phosphorylation site(s) catalyzed by different kinases.[68] While the stoichiometry of phosphorylation was not quantitatively measured, introducing phosphomimetic mutations into recombinant RGS9-1 reduced its GAP activity, suggesting that this may be a modulation mechanism for phototransduction recovery, akin to what was more recently shown for GRK1 by PKA.[69,70] While reversible protein phosphorylation is a common way of modulating enzyme activity in biochemistry, the extent to which it is used to modulate phototransduction was not understood to its full extent. This is another fertile area of discovery.

6. CONCLUSIONS: EMERGING FUNCTIONS OF RGS PROTEINS IN THE VISUAL SYSTEM

RGS9-1 belongs to the R7 subfamily of RGS proteins, which include three additional structurally similar members, namely RGS6, RGS7, and RGS11.[27] The four R7 RGS proteins contain the G protein gamma-like domain and use it to interact with the fifth member of the Gβ gene family Gβ5.[71] Gβ5 is peculiarly spliced in photoreceptor and contains an extra N-terminal exon.[50] The resulting protein, Gβ5-L, forms complex with

RGS9-1 and R9AP and function as the transducin GAP. Outside of photoreceptors and in the retina, Gβ5 is expressed in the so-called short form, Gβ5-S,[72] and interacts exclusively with RGS6, RGS7, and RGS11.[73] The redundant functions of Gβ5-S/RGS7 and Gβ5-S/RGS11 as the GAP for Goα in the metabotropic glutamate receptor 6 pathway at dendritic tips of depolarizing bipolar cells (DBCs) have been well-documented.[36,73,74] DBCs in mice lacking RGS7 and RGS11 possess undetectable Gβ5-S staining and have very poor light responses.[75] As a result, their ERG recordings are identical to those of the Gβ5$^{-/-}$ mice in that they both lack the characteristic ERG B-waves.[73,76] In addition, DBC dendritic tips are conspicuously stunted in Gβ5$^{-/-}$ mice and in one strain of the RGS7 and RGS11 double knockout mice,[73,76] but not in another double knockout strain where the *RGS7* gene was targeted differently.[75] The controversy may be worth solving for reasons mentioned previously. Perhaps more importantly, in the spirit of discovery, is that other than photoreceptors and DBCs, virtually nothing is known about RGS proteins in the rest of the visual system despite their abundant presence. Some RGS proteins are expressed early during development and thus may even have additional roles than merely "GAPing" heterotrimeric G-proteins. Finally, in intrinsically photosensitive retinal ganglion cells,[77] there exists another G-protein-mediated phototransduction pathway more similar to those in invertebrate ommatidia than in the one discussed above. This phototransduction pathway is initiated by melanopsin, used mainly for nonimage forming vision, and proceeds supposedly through Gq and with a much slower kinetics than the one in rod or cone.[78] Simple questions like which Gq family protein(s) or whether any RGS protein is involved in its signaling remain unanswered. While the role of RGS9-1/Gβ5-L/R9AP in rod phototransduction is understood, why and how they are expressed (or maintained) in higher level in cone or whether cone phototransduction is likewise rate-limited by transducin deactivation is presently unclear. A simultaneous comparative examination of both rods and cones will provide valuable insights. Future research efforts may thus benefit from a focused approach in the retina, due to its approachability and available anatomical and neurochemical details and genetic resources.

REFERENCES

1. Papermaster DS. Preparation of retinal rod outer segments. *Methods Enzymol.* 1982;81:48–52.
2. Raport CJ, Lem J, Makino C, et al. Downregulation of cGMP phosphodiesterase induced by expression of GTPase-deficient cone transducin in mouse rod photoreceptors. *Invest Ophthalmol Vis Sci.* 1994;35(7):2932–2947.

3. Miki N, Keirns JJ, Marcus FR, Freeman J, Bitensky MW. Regulation of cyclic nucleotide concentrations in photoreceptors: an ATP-dependent stimulation of cyclic nucleotide phosphodiesterase by light. *Proc Natl Acad Sci USA*. 1973;70(12):3820–3824.
4. Hagins WA. The visual process: excitatory mechanisms in the primary receptor cells. *Annu Rev Biophys Bioeng*. 1972;1:131–158.
5. Fesenko EE, Kolesnikov SS, Lyubarsky AL. Induction by cyclic GMP of cationic conductance in plasma membrane of retinal rod outer segment. *Nature*. 1985;313:310–313.
6. Chen CK. The vertebrate phototransduction cascade: amplification and termination mechanisms. *Rev Physiol Biochem Pharmacol*. 2005;154:101–121.
7. Dowling JE, Ripps H. Effect of magnesium on horizontal cell activity in the skate retina. *Nature*. 1973;242(5393):101–103.
8. Fain GL, Dowling JE. Intracellular recordings from single rods and cones in the mudpuppy retina. *Science*. 1973;180(4091):1178–1181.
9. Baylor DA, Lamb TD, Yau KW. The membrane current of single rod outer segments. *J Physiol*. 1979;288:589–611.
10. Sather WA, Detwiler PB. Intracellular biochemical manipulation of phototransduction in detached rod outer segments. *Proc Natl Acad Sci USA*. 1987;84(24):9290–9294.
11. Lamb TD, Pugh Jr EN. A quantitative account of the activation steps involved in phototransduction in amphibian photoreceptors. *J Physiol*. 1992;449:719–758.
12. Hurley JB. Termination of photoreceptor responses. *Curr Opin Neurobiol*. 1994;4(4):481–487.
13. Kuhn H, Dreyer WJ. Light dependent phosphorylation of rhodopsin by ATP. *FEBS Lett*. 1972;20(1):1–6.
14. Wilden U, Wust E, Weyand I, Kuhn H. Rapid affinity purification of retinal arrestin (48 kDa protein) via its light-dependent binding to phosphorylated rhodopsin. *FEBS Lett*. 1986;207(2):292–295.
15. Chen J, Makino CL, Peachey NS, Baylor DA, Simon MI. Mechanisms of rhodopsin inactivation in vivo as revealed by a COOH-terminal truncation mutant. *Science*. 1995;267(5196):374–377.
16. Chen CK, Burns ME, Spencer M, et al. Abnormal photoresponses and light-induced apoptosis in rods lacking rhodopsin kinase. *Proc Natl Acad Sci USA*. 1999;96(7):3718–3722.
17. Xu J, Dodd RL, Makino CL, Simon MI, Baylor DA, Chen J. Prolonged photoresponses in transgenic mouse rods lacking arrestin. *Nature*. 1997;389(6650):505–509.
18. Koch KW, Stryer L. Highly cooperative feedback control of retinal rod guanylate cyclase by calcium ions. *Nature*. 1988;334(6177):64–66.
19. Yau KW, Haynes LW, Nakatani K. Study of the roles of calcium and cyclic GMP in visual transduction. *Neurosci Res Suppl*. 1987;6:S45–S53.
20. Dizhoor AM, Olshevskaya EV, Henzel WJ, et al. Cloning, sequencing, and expression of a 24-kDa Ca(2+)-binding protein activating photoreceptor guanylyl cyclase. *J Biol Chem*. 1995;270(42):25200–25206.
21. Gorczyca WA, Polans AS, Surgucheva IG, Subbaraya I, Baehr W, Palczewski K. Guanylyl cyclase activating protein. A calcium-sensitive regulator of phototransduction. *J Biol Chem*. 1995;270(37):22029–22036.
22. Arshavsky VY, Wensel TG. Timing is everything: GTPase regulation in phototransduction. *Invest Ophthalmol Vis Sci*. 2013;54(12):7725–7733.
23. Angleson JK, Wensel TG. A GTPase-accelerating factor for transducin, distinct from its effector cGMP phosphodiesterase, in rod outer segment membranes. *Neuron*. 1993;11(5):939–949.
24. Arshavsky V, Bownds MD. Regulation of deactivation of photoreceptor G protein by its target enzyme and cGMP. *Nature*. 1992;357(6377):416–417.

25. Dohlman HG, Apaniesk D, Chen Y, Song J, Nusskern D. Inhibition of G-protein signaling by dominant gain-of-function mutations in Sst2p, a pheromone desensitization factor in *Saccharomyces cerevisiae*. *Mol Cell Biol.* 1995;15(7):3635–3643.
26. Koelle MR, Horvitz HR. EGL-10 regulates G protein signaling in the *C. elegans* nervous system and shares a conserved domain with many mammalian proteins. *Cell.* 1996;84(1):115–125.
27. Ross EM, Wilkie TM. GTPase-activating proteins for heterotrimeric G proteins: regulators of G protein signaling (RGS) and RGS-like proteins. *Annu Rev Biochem.* 2000;69:795–827.
28. Wieland T, Chen CK. Regulators of G-protein signalling: a novel protein family involved in timely deactivation and desensitization of signalling via heterotrimeric G proteins. *Naunyn Schmiedebergs Arch Pharmacol.* 1999;360(1):14–26.
29. Chen CK, Wieland T, Simon MI. RGS-r, a retinal specific RGS protein, binds an intermediate conformation of transducin and enhances recycling. *Proc Natl Acad Sci USA.* 1996;93(23):12885–12889.
30. He W, Cowan CW, Wensel TG. RGS9, a GTPase accelerator for phototransduction. *Neuron.* 1998;20(1):95–102.
31. Natochin M, Granovsky AE, Artemyev NO. Regulation of transducin GTPase activity by human retinal RGS. *J Biol Chem.* 1997;272(28):17444–17449.
32. Faurobert E, Hurley JB. The core domain of a new retina specific RGS protein stimulates the GTPase activity of transducin in vitro. *Proc Natl Acad Sci USA.* 1997;94(7):2945–2950.
33. Nekrasova ER, Berman DM, Rustandi RR, Hamm HE, Gilman AG, Arshavsky VY. Activation of transducin guanosine triphosphatase by two proteins of the RGS family. *Biochemistry.* 1997;36(25):7638–7643.
34. Snow BE, Antonio L, Suggs S, Siderovski DP. Cloning of a retinally abundant regulator of G-protein signaling (RGS-r/RGS16): genomic structure and chromosomal localization of the human gene. *Gene.* 1998;206(2):247–253.
35. Dhingra A, Faurobert E, Dascal N, Sterling P, Vardi N. A retinal-specific regulator of G-protein signaling interacts with Galpha(o) and accelerates an expressed metabotropic glutamate receptor 6 cascade. *J Neurosci.* 2004;24(25):5684–5693.
36. Chen FS, Shim H, Morhardt D, et al. Functional redundancy of R7 RGS proteins in ON-bipolar cell dendrites. *Invest Ophthalmol Vis Sci.* 2010;51(2):686–693.
37. Wieland T, Chen CK, Simon MI. The retinal specific protein RGS-r competes with the gamma subunit of cGMP phosphodiesterase for the alpha subunit of transducin and facilitates signal termination. *J Biol Chem.* 1997;272(14):8853–8856.
38. Chen CK, Burns ME, He W, Wensel TG, Baylor DA, Simon MI. Slowed recovery of rod photoresponse in mice lacking the GTPase accelerating protein RGS9-1. *Nature.* 2000;403(6769):557–560.
39. Lyubarsky AL, Chen C, Simon MI, Pugh Jr EN. Mice lacking G-protein receptor kinase 1 have profoundly slowed recovery of cone-driven retinal responses. *J Neurosci.* 2000;20(6):2209–2217.
40. Nishiguchi KM, Sandberg MA, Kooijman AC, et al. Defects in RGS9 or its anchor protein R9AP in patients with slow photoreceptor deactivation. *Nature.* 2004;427(6969):75–78.
41. Cheng JY, Luu CD, Yong VH, Mathur R, Aung T, Vithana EN. Bradyopsia in an Asian man. *Arch Ophthalmol.* 2007;125(8):1138–1140.
42. Hartong DT, Pott JW, Kooijman AC. Six patients with bradyopsia (slow vision): clinical features and course of the disease. *Ophthalmology.* 2007;114(12):2323–2331.
43. Michaelides M, Li Z, Rana NA, et al. Novel mutations and electrophysiologic findings in RGS9- and R9AP-associated retinal dysfunction (Bradyopsia). *Ophthalmology.* 2010;117(1), 120.e1–127.e1.

44. Gold SJ, Ni YG, Dohlman HG, Nestler EJ. Regulators of G-protein signaling (RGS) proteins: region-specific expression of nine subtypes in rat brain. *J Neurosci.* 1997;17(20):8024–8037.
45. Zhang K, Howes KA, He W, et al. Structure, alternative splicing, and expression of the human RGS9 gene. *Gene.* 1999;240(1):23–34.
46. Martemyanov KA, Arshavsky VY. Biology and functions of the RGS9 isoforms. *Prog Mol Biol Transl Sci.* 2009;86:205–227.
47. Rinner O, Makhankov YV, Biehlmaier O, Neuhauss SC. Knockdown of cone-specific kinase GRK7 in larval zebrafish leads to impaired cone response recovery and delayed dark adaptation. *Neuron.* 2005;47(2):231–242.
48. Lem J, Makino CL. Phototransduction in transgenic mice. *Curr Opin Neurobiol.* 1996;6(4):453–458.
49. Nikonov SS, Brown BM, Davis JA, et al. Mouse cones require an arrestin for normal inactivation of phototransduction. *Neuron.* 2008;59(3):462–474.
50. Watson AJ, Aragay AM, Slepak VZ, Simon MI. A novel form of the G protein beta subunit Gbeta5 is specifically expressed in the vertebrate retina. *J Biol Chem.* 1996;271(45):28154–28160.
51. Hu G, Wensel TG. R9AP, a membrane anchor for the photoreceptor GTPase accelerating protein, RGS9-1. *Proc Natl Acad Sci USA.* 2002;99(15):9755–9760.
52. Makino ER, Handy JW, Li T, Arshavsky VY. The GTPase activating factor for transducin in rod photoreceptors is the complex between RGS9 and type 5 G protein beta subunit. *Proc Natl Acad Sci USA.* 1999;96(5):1947–1952.
53. Keresztes G, Martemyanov KA, Krispel CM, et al. Absence of the RGS9.Gbeta5 GTPase-activating complex in photoreceptors of the R9AP knockout mouse. *J Biol Chem.* 2004;279(3):1581–1584.
54. Krispel CM, Chen CK, Simon MI, Burns ME. Prolonged photoresponses and defective adaptation in rods of Gbeta5-/- mice. *J Neurosci.* 2003;23(18):6965–6971.
55. Cowan CW, Fariss RN, Sokal I, Palczewski K, Wensel TG. High expression levels in cones of RGS9, the predominant GTPase accelerating protein of rods. *Proc Natl Acad Sci USA.* 1998;95(9):5351–5356.
56. Zhang X, Wensel TG, Kraft TW. GTPase regulators and photoresponses in cones of the eastern chipmunk. *J Neurosci.* 2003;23(4):1287–1297.
57. Chen CK, Eversole-Cire P, Zhang H, et al. Instability of GGL domain-containing RGS proteins in mice lacking the G protein beta-subunit Gbeta5. *Proc Natl Acad Sci USA.* 2003;100(11):6604–6609.
58. Krispel CM, Chen D, Melling N, et al. RGS expression rate-limits recovery of rod photoresponses. *Neuron.* 2006;51(4):409–416.
59. Pugh Jr EN. RGS expression level precisely regulates the duration of rod photoresponses. *Neuron.* 2006;51(4):391–393.
60. Slepak VZ, Hurley JB. Mechanism of light-induced translocation of arrestin and transducin in photoreceptors: interaction-restricted diffusion. *IUBMB Life.* 2008;60(1):2–9.
61. Sokolov M, Lyubarsky AL, Strissel KJ, et al. Massive light-driven translocation of transducin between the two major compartments of rod cells: a novel mechanism of light adaptation. *Neuron.* 2002;34(1):95–106.
62. Strissel KJ, Lishko PV, Trieu LH, Kennedy MJ, Hurley JB, Arshavsky VY. Recoverin undergoes light-dependent intracellular translocation in rod photoreceptors. *J Biol Chem.* 2005;280(32):29250–29255.
63. Chen CK, Inglese J, Lefkowitz RJ, Hurley JB. Ca(2+)-dependent interaction of recoverin with rhodopsin kinase. *J Biol Chem.* 1995;270(30):18060–18066.
64. Tian M, Zallocchi M, Wang W, et al. Light-induced translocation of RGS9-1 and Gbeta5L in mouse rod photoreceptors. *PLoS One.* 2013;8(3):e58832.

65. Hu G, Jang GF, Cowan CW, Wensel TG, Palczewski K. Phosphorylation of RGS9-1 by an endogenous protein kinase in rod outer segments. *J Biol Chem*. 2001;276(25):22287–22295.
66. Sokal I, Hu G, Liang Y, Mao M, Wensel TG, Palczewski K. Identification of protein kinase C isozymes responsible for the phosphorylation of photoreceptor-specific RGS9-1 at Ser475. *J Biol Chem*. 2003;278(10):8316–8325.
67. Whitlock GG, Lamb TD. Variability in the time course of single photon responses from toad rods: termination of rhodopsin's activity. *Neuron*. 1999;23(2):337–351.
68. Balasubramanian N, Levay K, Keren-Raifman T, Faurobert E, Slepak VZ. Phosphorylation of the regulator of G protein signaling RGS9-1 by protein kinase A is a potential mechanism of light- and Ca2+-mediated regulation of G protein function in photoreceptors. *Biochemistry*. 2001;40(42):12619–12627.
69. Horner TJ, Osawa S, Schaller MD, Weiss ER. Phosphorylation of GRK1 and GRK7 by cAMP-dependent protein kinase attenuates their enzymatic activities. *J Biol Chem*. 2005;280(31):28241–28250.
70. Osawa S, Jo R, Xiong Y, et al. Phosphorylation of G protein-coupled receptor kinase 1 (GRK1) is regulated by light but independent of phototransduction in rod photoreceptors. *J Biol Chem*. 2011;286(23):20923–20929.
71. Snow BE, Krumins AM, Brothers GM, et al. A G protein gamma subunit-like domain shared between RGS11 and other RGS proteins specifies binding to Gbeta5 subunits. *Proc Natl Acad Sci USA*. 1998;95(22):13307–13312.
72. Watson AJ, Katz A, Simon MI. A fifth member of the mammalian G-protein beta-subunit family. Expression in brain and activation of the beta 2 isotype of phospholipase C. *J Biol Chem*. 1994;269(35):22150–22156.
73. Shim H, Wang CT, Chen YL, et al. Defective retinal depolarizing bipolar cells in regulators of G protein signaling (RGS) 7 and 11 double null mice. *J Biol Chem*. 2012;287(18):14873–14879.
74. Mojumder DK, Qian Y, Wensel TG. Two R7 regulator of G-protein signaling proteins shape retinal bipolar cell signaling. *J Neurosci*. 2009;29(24):7753–7765.
75. Cao Y, Pahlberg J, Sarria I, Kamasawa N, Sampath AP, Martemyanov KA. Regulators of G protein signaling RGS7 and RGS11 determine the onset of the light response in ON bipolar neurons. *Proc Natl Acad Sci USA*. 2012;109(20):7905–7910.
76. Rao A, Dallman R, Henderson S, Chen CK. Gbeta5 is required for normal light responses and morphology of retinal ON-bipolar cells. *J Neurosci*. 2007;27(51):14199–14204.
77. Berson DM, Dunn FA, Takao M. Phototransduction by retinal ganglion cells that set the circadian clock. *Science*. 2002;295(5557):1070–1073.
78. Xue T, Do MT, Riccio A, et al. Melanopsin signalling in mammalian iris and retina. *Nature*. 2011;479(7371):67–73.

CHAPTER FOUR

Role of Regulators of G Protein Signaling Proteins in Bone Physiology and Pathophysiology

Joel Jules*, Shuying Yang†,‡, Wei Chen*, Yi-Ping Li*,[1]
*Department of Pathology, School of Medicine, University of Alabama at Birmingham, Birmingham, Alabama, USA
†Department of Oral Biology, School of Dental Medicine, University at Buffalo, The State University of New York, Buffalo, New York, USA
‡Developmental Genomics Group, New York State Center of Excellence in Bioinformatics and Life Sciences, University at Buffalo, The State University of New York, Buffalo, New York, USA
[1]Corresponding author: e-mail address: ypli@uab.edu

Contents

Progress in Molecular Biology and Translational Science, Volume 133
ISSN 1877-1173
http://dx.doi.org/10.1016/bs.pmbts.2015.02.002

Abstract

Regulators of G protein signaling (RGS) proteins enhance the intrinsic GTPase activity of α subunits of the heterotrimeric G protein complex of G protein-coupled receptors (GPCRs) and thereby inactivate signal transduction initiated by GPCRs. The RGS family consists of nearly 37 members with a conserved RGS homology domain which is critical for their GTPase accelerating activity. RGS proteins are expressed in most tissues, including heart, lung, brain, kidney, and bone and play essential roles in many physiological and pathological processes. In skeletal development and bone homeostasis as well as in many bone disorders, RGS proteins control the functions of various GPCRs, including the parathyroid hormone receptor type 1 and calcium-sensing receptor and also regulate various critical signaling pathways, such as Wnt and calcium oscillations. This chapter will discuss the current findings on the roles of RGS proteins in regulating signaling of key GPCRs in skeletal development and bone homeostasis. We also will examine the current updates of RGS proteins' regulation of calcium oscillations in bone physiology and highlight the roles of RGS proteins in selected bone pathological disorders. Despite the recent advances in bone and mineral research, RGS proteins remain understudied in the skeletal system. Further understanding of the roles of RGS proteins in bone should not only provide great insights into the molecular basis of various bone diseases but also generate great therapeutic drug targets for many bone diseases.

1. INTRODUCTION TO BONE

Bone is a connective tissue that plays crucial roles in mineral storage and homeostasis, organ support, and locomotion.[1,2] Bone is composed of organic and inorganic materials. The inorganic component of bone is primarily constituted of hydroxyapatite, $[Ca_3(PO4)_2]_3\ Ca(OH)_2$. The organic portion of bone consists mainly of type I collagen, a triple-helical molecule containing three polypeptide chains of amino acids that are each cross-linked by hydrogen bonds. The remaining organic constituent of bone contains various noncollagenous proteins, including hormones, growth factors, and cytokines.[3–5] Bone formation occurs through two key mechanisms: intramembranous and endochondral ossification.[4,6] Intramembranous mineralization gives rise to the cranial vault, some facial bones, parts of the mandible and clavicles, whereas endochondral ossification is responsible for the other bones of the skeleton. During intramembranous bone formation, undifferentiated mesenchymal cells give rise to osteoprogenitor cells which then differentiate into mature osteoblasts.[3] During endochondral ossification, mesenchymal stem cells are first condensed to form a template of cartilage that is ultimately replaced by bone. Throughout adult life, bone is constantly remodeling through the synchronized and balanced activities of the osteoclasts, which derive from the hematopoietic stem cell lineage and are responsible for resorbing bone, and the osteoblasts, the bone-forming cells of the mesenchymal stem cell origin.[7–9] Disruption which shifts the balance in the favor of osteoclasts can affect bone mass and causes many pathological bone disorders, including osteoporosis, periodontitis, endodontitis, and rheumatoid arthritis.[10–14] Likewise, enhanced function of osteoblasts over osteoclasts can cause osteopetrosis.

2. THE GPCR–G PROTEIN–RGS SIGNALING PATHWAY

The activities of osteoclasts and osteoblasts are highly controlled by autocrine, paracrine, and endocrine factors from the external environment to ensure the systemic balance of calcium–phosphate metabolism while maintaining bone homeostasis.[5] External stimuli can affect bone cells by binding to their receptors on the cell membranes and thereby trigger signals within the cells. G protein-coupled receptors (GPCRs) are an example of such receptors.[15,16] GPCRs, also called seven-transmembrane domain receptors, are a large family of protein receptors. These receptors sense a variety of extracellular stimuli from growth factors, cytokines, hormones, neurotransmitters, light, to phospholipids to affect various cellular processes, such as cell proliferation, differentiation, activity, and apoptosis.[17] Specifically, GPCR activation transduces intracellular signals through a heterotrimeric G protein complex which can then direct the signals to downstream effectors for specific outcomes (Fig. 1). GPCRs regulate many physiological events and, consequently, have been exploited therapeutically in many disease states, including diabetes, various cancers as well as bone, neurological, diseases, blood, heart, and kidney diseases.[18–21] Nearly 40% of the current drugs on the market target GPCRs.

2.1 The G Protein Complex

GPCRs are linked to a heterotrimeric G protein complex at its cytoplasmic domain[22] (Fig. 1). The G protein heterotrimeric complexes consist of three different subunits: Gα (33–55 kDa), Gβ (~35 kDa), and Gγ (~15 kDa).[21,22] In humans, the Gα subunits are encoded by 16 genes and are divided into 4 subgroups ($G\alpha_s$, $G\alpha_{o/i}$, $G\alpha_{q/11}$, and $G\alpha_{12/13}$) which are further divided into other subgroups. The Gβ subunits are encoded by 5 genes and are divided into 6 subgroups. There are 12 different Gγ subunits. Given that signaling transduced by numerous combinations of βγ subunits is not significantly

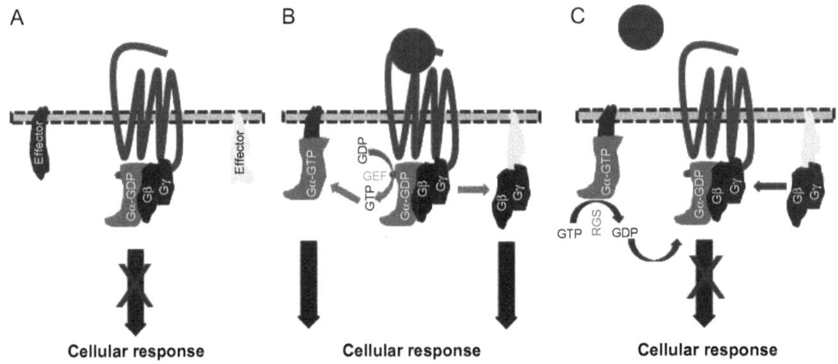

Figure 1 Schematic of GPCR–G protein–RGS activation and inactivation cycle. (A) GPCR inactivation in the absence of ligands. In absence of ligands, Gα is linked to GDP (Gα-GDP) and forms a heterotrimeric complex with the Gγ and Gβ subunits, preventing their liberation to activate effectors to induce cellular response. (B) Ligand-induced GPCR activation. In the presence of ligands, GPCR undergoes a conformational change leading to the activation of Gα through the exchange of GDP for GTP by the receptor GEF activity. Gα-GTP and Gβγ subunits dissociate to modulate the activity of specific effectors to mediate cellular responses. (C) GPCR signaling inactivation by RGS proteins. RGS proteins enhance the GTPase activity of Gα to hydrolyze GTP to GDP leading to the inactivation of the Gα protein and its subsequent reunification with Gβγ subunits to reform the inactive G protein heterotrimeric complex. This prevents the G proteins from continuing activating effectors for cellular response and thus terminates the cellular response. GDP, guanosine diphosphate; GTP, guanosine triphosphate; GEF, guanine nucleotide exchange factor. (See the color plate.)

different from one another, the G protein heterotrimeric complexes are mostly defined according to their Gα constituents. Upon its activation, a GPCR undergoes a conformational change in its cytoplasmic domain which leads to the activation of the Gα submit through the exchange of its bound GDP for GTP by the guanine nucleotide exchange factor (GEF) activity of GPCRs[16] (Fig. 1). The activated α subunit (Gα-GTP) then dissociates from the βγ submits to form two different G protein units which then transduce signaling through activation or inhibition of specific effectors to induce cellular responses.

Whereas $G\alpha_s$ can stimulate the adenylate cyclase (AC) pathway to activate the cyclic adenosine monophosphate (cAMP), activation of $G\alpha_i$ results in the inhibition of the cAMP pathway. $G\alpha_q$ activates the phospholipase C-β (PLCβ) pathway which cleaves phosphatidylinositol 4,5-biphosphate (PIP2) into inositol (1,4,5) triphosphate (IP3) and diacylgycerol (DAG), which can trigger an increase in intracellular Ca^{2+} levels. $G\alpha_{12/13}$ activates the Rho pathway which is involved in cell cytoskeleton biology. The Gβγ submits can activate many ion channels. Notably, GPCR signaling is terminated when the Gα subunits hydrolyses their bound GTP to GDP, via their intrinsic GTPase activity, to reunite with the Gβγ subunits to reform an inactive G protein heterodimer complex (Fig. 1). The kinetics of G protein signaling are controlled by a family of proteins, named regulators of G protein signaling (RGS), which bind to activated Gα and accelerate their intrinsic GTPase activities.[23,24] Therefore, RGS proteins are crucial for the rapid activation and inactivation of cellular responses that are initiated by GPCR stimulation.[25]

2.2 Introduction to RGS Proteins

The RGS protein family consists of at least 37 members in mammals and is divided into nine subfamilies, including RZ/A, R4/B, R7/C, R12/E, GEF/F, GRK/G, SNX/H, and D-AKAP2/I based on structural analysis[18] (Fig. 2). Nevertheless, all RGS proteins contain a highly conserved 120-amino acid RGS domain, which can enhance the GTPase activity of the

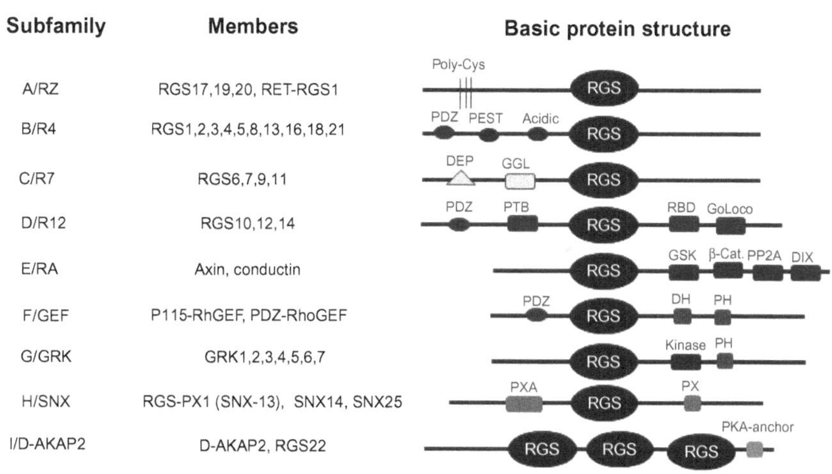

Figure 2 Subclassification of RGS family members. All members of the RGS family possess a common RGS domain but are subdivided into different subfamilies based on protein structures. β-Cat, β-catenin-binding; D-AKAP, dual-specificity A-kinase anchoring protein; DEP, dishevelled/EGL-10/pleckstrin; DH, double homology; DIX, dishevelled homology domain; GAIP, Gα interacting protein; GEF, guanine nucleotide exchange factor; GGL, Gγ-like; Goloco, $G\alpha_{i/o}$-Loco; GRK, G protein-coupled receptor kinase; GSK, glycogen synthase kinase 3β-binding; PDZ, PSD95/dlg/ZO-1/2; PEST, proline, glutamine, serine, threonine-rich; PH, pleckstrin homology; PP2A, protein phosphatase 2A; PTB, phosphotyrosine-binding; PX, phosphatidylinositol-binding; PXA, PX-associated; RBD, Ras-binding domain; SNX, sorting nexin. (See the color plate.)

Gα proteins.[26] The GTPase accelerating protein (GAP) of RGS proteins is mainly limited to $G\alpha_q$, $G\alpha_{i/o}$, and $G\alpha_{12/13}$ subunits. The functions of RGS proteins are essential for coordinating signaling output elicited by GPCRs. Indeed, RGS proteins can regulate G protein-mediated receptors, ion channels, and other signaling pathways and can also serve as key modulators in many physiological processes, including cardiovascular, respiratory, nervous and immune functions as well as many disease states, such as diabetes, cardiovascular diseases, cancers, inflammatory diseases, and neurological diseases.[27–31] RGS proteins are also critical for the skeletal system by regulating various signaling pathways, including Wnt, parathyroid hormone (PTH), and calcium-sensing receptor (CaSR) pathways and Ca^{2+} signaling.[32,33] Consistent with their critical roles in bone, it was shown that the expression of RGS proteins can be regulated by many bone factors, including proinflammatory cytokines, lipopolysaccharide (LPS), PTH and PTH-related peptide (PTHrP), and thyroid-stimulating hormone (TSH).[33–35] LPS can inhibit the expression of RGS2 in macrophages but increases the expression of RGS1 in dendritic cells and macrophages.[36–38] Likewise, interferon β can upregulate the expression of RGS1 in monocytes, B cells, and T cells and induce the expression of RGS2 and RGS16 in mononuclear leukocytes.[38,39] Moreover, PTH and PTHrP can induce RGS2 in osteoblasts. Injection of PTH in femoral metaphysical spongiosa of young male rats can increase the expression of RGS2 in a rapid but transient fashion. Moreover, the expression of RGS2 can also be induced by TSH stimulation.[35] As such, targeting RGS protein functions can attenuate the symptoms associated with many bone pathological disorders. Here, we summarized the current knowledge on the roles of RGS proteins in the bone-forming osteoblasts and bone-resorbing osteoclasts. We also reviewed the roles of RGS proteins in some key-skeletal signaling pathways under normal physiological and pathological states.

3. RGS PROTEINS IN OSTEOBLASTS

Osteoblasts are mononucleated cells that derive from the mesenchymal stem cell lineage in the bone marrow and are responsible for bone formation.[6] Osteoblasts induce bone mineralization by producing new bone called "osteoid." Bone formation by osteoblasts are influenced by many GPCRs, including PTH1R, frizzled (Fz), and CaSR,[21] which are critical for osteoblast differentiation and function. Transgenic mice bearing a constitutively active form of CaSR in mature osteoblasts via the osteocalcin promoter cause a decrease in bone volume and density in cancellous bone.[21,40] These effects result from enhanced osteoclast formation and activity triggered by an increase in the expression of receptor activator of nuclear factor κβ ligand (RANKL). Furthermore, CaSR can also promote osteoblast proliferation in part through the activation of the $G\alpha_q$-PLCβ pathway leading to increased Ca^{2+} oscillations.[16] Collectively, these studies demonstrate a role for GPCRs in osteoblast differentiation and suggest a role for RGS proteins in osteoblast biology. Further studies reported that overexpression of RGS2 or RGS4 and pretreatment with pertussis toxin can inhibit the activation of GPRC6A by extracellular cations.[41] However, RGS proteins have not been highly studied in osteoblasts. Here, we discussed the current understanding of the known roles of RGS2, RGS5, and axin in osteoblasts.

3.1 RGS2 in Osteoblasts

RGS2 regulates G protein-mediated functions in various tissues, including lung, brain, heart, and bone.[18] RGS2 is expressed in the rat metaphyseal and diaphyseal, mouse calvarial culture, and cultured osteoblasts and plays a critical role in the skeletal system. Nevertheless, RGS2 function in osteoblasts is a bit complex. RGS2 expression is upregulated by forskolin, PTH, and PTHrP in osteoblasts through the stimulation of the AC-cAMP pathway by $G\alpha_s$ activation,[42–45] which is critical for osteoblast differentiation. It was reported that a high level of RGS2 is required to attenuate signaling by GPCRs that promote Ca^{2+} mobilization via the activation of the $G\alpha_q$-PLCβ pathway β.[16,18] RGS2 can also be induced by ATP-induced activation of $G\alpha_q$ to attenuate PTH-induced activation of cAMP. Overall, the finding indicates that, at basal levels RGS2 does not regulate $G\alpha_q$ or $G\alpha_s$ signaling, but at higher levels, RGS2 may cross-desensitize both $G\alpha_s$ and $G\alpha_q$ in osteoblasts.

3.2 RGS5 in Osteoblasts

Recent findings have suggested that PTH-induced RGS5 expression may play a crucial role in cellular responses to extracellular calcium level through activation of CaSR.[46] RGS5 was reported to be markedly elevated in parathyroid adenomas, and mice deficient in RGS5 showed a reduced level of PTH plasma levels.[46–48] Notably, forced expression of RGS5 in cells expressing CaSR abrogates the Ca^{2+}-induced IP3 induction by CaSR. These results indicate that RGS5 can act as a physiological negative regulator of CaSR in the parathyroid gland. Unfortunately, our understanding of the role of RGS5 in organ physiology is mostly unknown because of a lack of RGS5-targeted mouse models.[18] Nonetheless, RGS5 is also expressed in smooth muscle and heart. A novel splice variant of RGS5, RGS5S, which lacks 104 amino acids has recently been cloned and reported to be expressed in the human brain, skeletal system, eye, and small intestine.[49]

3.3 Axin in Osteoblasts

Axin, a member of E/RA subfamily of the RGS protein family, is an important regulator of skeletal homeostasis (Fig. 3).[50–53] Axin serves as molecular scaffold for the β-catenin destruction complex by interacting with other proteins, such as protein phosphatase 2A (PP2A), adenomatosis polyposis coli (APC), glycogen synthase kinase 3 (GSK3), and casein kinase 1 (CK1), thus functioning as a negative regulator of the canonical Wnt–β-catenin signaling pathway.[32] Mice deficient in the axin gene display a significant increase in bone density stemming from an increase in osteoblast

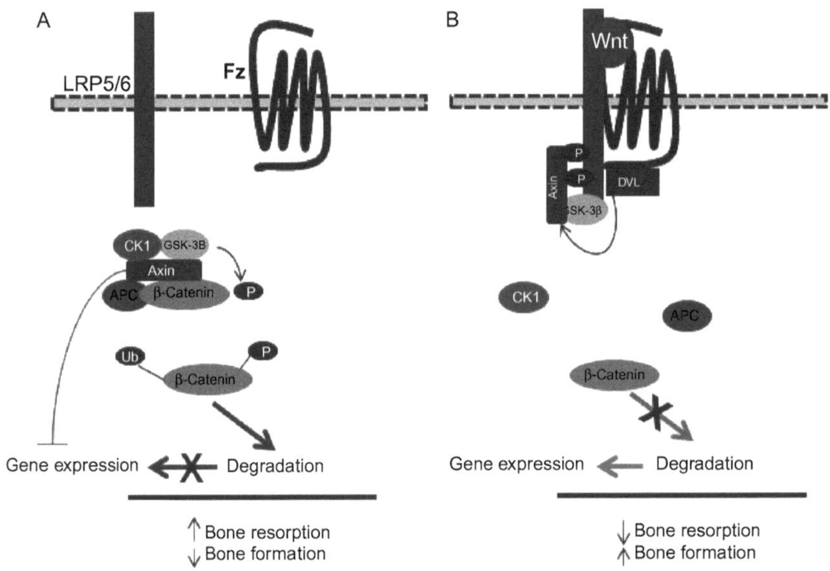

Figure 3 Role of axin in Wnt signaling in bone. (A) In the absence of Wnt ligands, β-catenin is constructively targeted for degradation via a multiprotein destruction complex. β-Catenin is phosphorylated by GSK-3β and then targeted for ubiquitination (Ub) and proteosomal degradation. Axin plays a key role in this process of β-catenin degradation and thus functions as an inhibitor of the canonical Wnt-β pathway. (B) Binding of canonical Wnt ligands to their co-receptors Fz and LRP5/6 prevent β-catenin phosphorylation by GSK-3β, which in turn prevent its degradation and allows it to move to the nucleus to mediate gene expression. (See the color plate.)

differentiation and a decrease in osteoclast differentiation.[50] Axin can also negatively regulate bone mass through the β-catenin-bone morphogenetic proteins 2 and 4 (BMP2/4)–osterix (Osx) signaling pathway in osteoblasts.[16] Axin has also been shown to interact with N-cadherin to mediate β-catenin degradation leading to decrease osteoblast differentiation.[54] Consistent with that posture, transgenic mice expressing N-cadherin under the control of the Col1 2.3 promoter display an increase in β-catenin degradation leading to a decrease in osteoblasts and bone mass.[54,55] Notably, axin may also control the Wnt–β-catenin signaling pathway through actions on Gα proteins. Whereas $G\alpha_s$ proteins bearing activating mutations can interact with axin to enhance β-catenin activation, a lack of $G\alpha_s$ causes an increase in β-catenin degradation which affects bone mass. All together, these results demonstrate a crucial role for axin as an important negative regulator of bone formation through inhibition of the Wnt–β-catenin pathway.

4. RGS PROTEINS IN OSTEOCLASTS

Osteoclasts play essential roles in skeletal development and bone homeostasis.[7] Enhanced osteoclast formation and/or activity are responsible for many disorders of skeletal deficiency.[10] Osteoclastogenesis requires two key cytokines: RANKL and macrophage colony-stimulation factor (M-CSF). Whereas M-CSF maintains the proliferation of osteoclast precursors, RANKL is responsible for the differentiation of these precursors into mature osteoclasts. RANKL functions by activating its receptor, RANK, to induce the expression of osteoclast marker genes, including the nuclear factor of activated T cells, c1 (NFATc1), a master regulator of osteoclast differentiation.[56,57] Osteoclastogenesis also requires the activation of costimulatory signals from immunoreceptor tyrosine-based activation motif-containing (ITAM-containing) receptors, mainly the DNAX-activating protein 12 (DAP12) and Fc receptor common γ (FcRγ) subunits, in osteoclast precursors.[58] Activation of DAP12 and FcRγ triggers the activation of PLC which can maintain intracellular calcium $[Ca^{2+}]i$ oscillations, which is critical for the activation of NFATc1.[58–60] Consequently, recombinant recognition sequence binding protein at the Jκ site (RBP-J), one of the key inhibitors of osteoclastogenesis, functions by blocking PLC-induced $[Ca^{2+}]i$ oscillations in osteoclast precursors.[61,62]

In line with the requirement for Ca^{2+}, osteoclasts have an active Ca^{2+} regulatory system, including a membrane Ca^{2+}-ATPase and endoplasmic Ca^{2+}-ATPase as well as receptors, such as ovarian cancer G protein-coupled receptor (ORG1) and N-type Ca^{2+} channel, to regulate [Ca^{2+}]i levels.[16] Also, the basomembrane of osteoclasts is very sensitive to elevated Ca^{2+} levels.[16,21,63] Despite the essential requirement for calcium, which can be activated by the GPCR–$G\alpha_q$–PLC signaling and the findings that osteoclasts expressed many GPCRs, such as calcitonin receptor, ORG1, and CaSR, the roles of GPCR–G proteins–RGS proteins in regulating osteoclast formation and/or activity remain understudied.[3,21,64,65] Below, we reviewed the reported roles of RGS10, RGS12, and RGS18 in osteoclasts.

4.1 RGS10 in Osteoclasts

RGS10 is expressed by many tissues and cells, including brain, testis, atrial monocytes, B lymphocytes, dendritic cells, and osteoclasts.[16] RGS10 has two key isoforms: RGS10A and RGS10B. We found that, through a differential screening of a human osteosarcoma cDNA library, RGS10A, but not RGS10B, is specifically expressed by human osteoclasts.[66] Consequently, we demonstrated that RGS10A is upregulated by RANKL in osteoclast precursors and highly expressed in murine osteoclast-like cells. RGS10A silencing by RNA interference blocks [Ca^{2+}]i oscillations, the expression of NFATc1, and osteoclast differentiation in bone marrow cells and the osteoclast cell lines. Most importantly, we also showed that mice deficient in the *RGS10* gene (*RGS10*$^{-/-}$ mice) display a severe osteopetrotic phenotype as a result of defective osteoclast formation.[67] RGS10 overexpression markedly enhances the sensitivity of osteoclast differentiation to RANKL stimulation, restores [Ca^{2+}]i oscillations, and induces NFATc1 expression. Interestingly, forced expression of NFATc1 could also rescue osteoclastogenesis in RGS10$^{-/-}$ bone marrow cells, indicating that RGS10 functions upstream of NFATc1. Hence, we revealed that RGS10 competitively interacts with Ca^{2+}/calmodulin (CaM) and IP3 in a calcium-dependent manner to mediate PLC activation and [Ca^{2+}]i oscillations for the ultimate activation of NFATc1.[3] Hence, our studies have established RGS10 as a crucial regulator of osteoclast differentiation and bone homeostasis.

4.2 RGS12 in Osteoclasts

RGS12 is expressed in many tissues and cells, including lung, spleen, ovary, kidney, prostate, brain, testis, and osteoclasts.[16] RGS12 is the largest member of the RGS protein family (Fig. 2). It contains many domains, including the RGS domain, a GoLoco motif which bears guanine nucleotide dissociation inhibitor (GDI) activity toward $G\alpha_i$ subunits, and a Ras-binding domain which provides additional functions to RGS12. Also, RGS12 possesses a PDZ and PTB domains which bind to the C terminus of GPCRs and interacts with GPCR chemokine receptors, respectively.[16] Given its multiple domains, RGS12 may regulate numerous signaling pathways. In fact, it has been shown that RGS12 can interact with certain types of calcium channels to control calcium oscillations in many cell types. In line with this finding, we found that RGS12 is highly induced by RANKL in osteoclast precursors and plays a crucial role in osteoclastogenesis.[68] RGS12 knockdown by RNA interference impairs osteoclastogenesis by affecting PLC activation which in turn impacts Ca^{2+} oscillations and the subsequent NFATc1 activation. In addressing this issue, we reported that RANKL can activate the N-type calcium channels to increase $[Ca^{2+}]i$ levels.[69–71] Notably, we revealed that RGS12 can interact with N-type calcium channel to regulate its activity. Consistently, targeted deletion of the RGS12 in the hematopoietic cell lineage results in a severe osteopetrotic phenotype from impaired osteoclastogenesis stemming from defective Ca^{2+} signaling. Whereas the defect in osteoclastogenesis from $RGS12^{-/-}$ osteoclast precursor cells could be rescued by overexpressing either RGS10 or RGS12, those in $RGS10^{-/-}$ bone marrow cells can only be rescued by RGS10. Given that RGS10 can activate calcium oscillations through induction of the IP3 pathway while RGS12 can promote calcium influxes through a specific calcium channel, these findings suggest that $[Ca^{2+}]i$ oscillations may be the most effective way to maintain $[Ca^{2+}]i$ levels in osteoclasts. Hence, our studies reveal critical roles for RGS10 and RGS12 in regulating calcium levels for osteoclast differentiation.

4.3 RGS18 in Osteoclasts

RGS18 is predominantly expressed by selected bone marrow cells, such as granulocytes, monocytes, and platelets but not by lymphocytes and erythrocytes.[18] Investigation of the role of RGS18 in osteoclasts revealed that RANKL attenuates RGS18 expression in the osteoclast precursor cell line and in the primary bone marrow-derived osteoclast precursor monocytes during osteoclastogenesis.[72] And, RGS18 depletion by RNA interference can promote osteoclastogenesis by RANKL stimulation. However, deletion of RGS18 is unable to promote osteoclast differentiation in the absence of RANKL stimulation, indicating that the RGS18 deficiency requires RANKL to stimulate osteoclastognesis. Previous reports indicated a role for OGR1 in osteoclastogenesis which stimulates the $G\alpha_q$–PLC pathway leading to the activation of NFATc1.[73–76] Interestingly, RGS18 inhibits osteoclastogenesis by acidosis-induced ORGR1 activation by affecting PLC-induced NFATc1 activation because the enhanced osteoclastogenesis from RGS18$^{-/-}$ cells can be revered by treatment with anti-ORG1 blocking antibody. Moreover, whereas overexpression of RGS18 can inhibit the NFATc1 activation triggered by ORG1 activation, RGS18 deletion can enhance NFATc1 activation in these conditions. These findings indicate that RGS18 can negatively regulate osteoclastogenesis induced by extracellular acidosis via activation of ORG1.

5. GPCR–RGS PROTEINS SIGNALING IN SKELETAL PHYSIOLOGY

5.1 RGS Proteins and PTH/PTHrP Signaling in Bone

PTH and PTHrP play crucial roles in bone homeostasis through their actions on osteoblasts via the PTH1R, a member of the GPCR family (Fig. 4).[77–80] Ligand binding on PTHR1 triggers the activation of the $G\alpha_q$–PLC and $G\alpha_s$–AC signaling pathways leading to $[Ca^{2+}]$i mobilizations from IP3 and increases in cAMP levels, respectively.[32] Increased $[Ca^{2+}]$ and cAMP levels are critical for osteoblast differentiation and function. Notably, the inactivation of PTHR1 signaling, mediated by specific RGS proteins, is vital to maintain the rapid turnover of cellular responses. Many studies have demonstrated a role for RGS2 in inactivating PTH1R signaling through the evidence that forced expression of RGS2 can attenuate PTH-induced cAMP production in osteoblasts.[16,81] Consistent with the role of RGS2 in inactivating PTHR1 signaling, PTH has been shown to be able to upregulate RGS2 expression both *in vivo* and *in vitro*.[81,82] Notably, PTH-induced RGS2 expression can also be inhibited by 1,25-$(OH)_2$-D_3 in a dose-dependent manner.[81] Moreover, while pretreatment with vitamin D can promote PTH-induced RGS2 expression, pretreatment with glucocorticoids such as dexamethasone can inhibit upregulation of RGS2 induced by PTH stimulation.[82] Hence, PTH-induced RGS2 upregulation may play an important mechanism by which PTH controls its activation.

5.2 RGS Proteins and Wnt Signaling in Bone

The canonical Wnt–β-catenin pathway is critical for skeletal development and bone homeostasis by its central function in osteoblast differentiation.[32] Wnt signaling is induced by the binding of Wnt ligands to the dual receptor complex consisted of Fz, a member of the GPCR family, and either

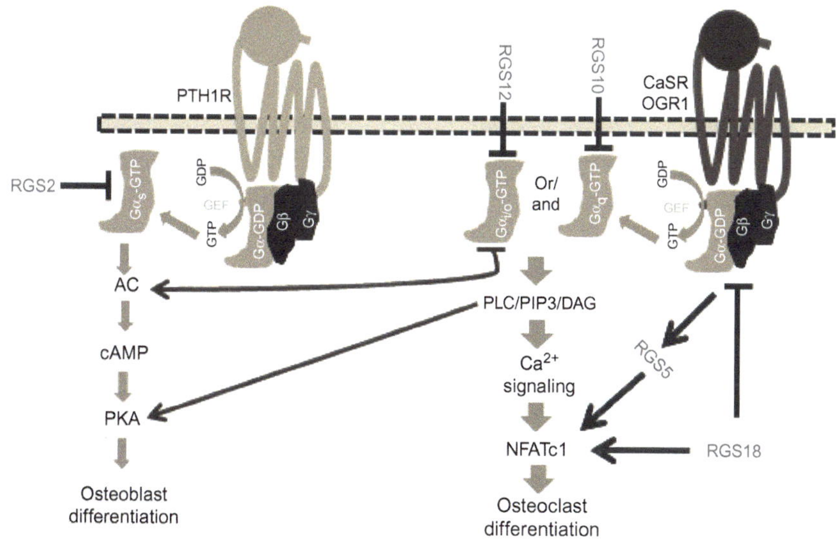

Figure 4 Role of RGS proteins in selected GPCR signaling in bone. In osteoblasts, PTH1R activation upon binding of PTH or PTHrP triggers the activation of the $G\alpha_s$–AC–cAMP signaling pathway which ultimately promotes osteoblast differentiation. Activation of PTH1R can induce the expression of many RGS proteins, including RGS2 and RGS5. RGS2 can target the cAMP signaling to regulate PTH1R signaling, and RGS5 is likely to regulate osteoblast differentiation by controlling PTH levels. In osteoclasts, activation of GPCRs, such as CasR and OGR1, can activate $G\alpha_q$ or $G\alpha_{i/o}$ to activate the PLC–Ca^{2+} signaling pathway for NFATc1 activation or $G\alpha_s$ to activate the Ac–cAMP pathway to promote osteoblast differentiation and function. RGS10, RGS12, and RGS18 can regulate osteoclast differentiation. (See the color plate.)

lipoprotein receptor-related protein 5 (LRP5) or LRP6 (Fig. 3). In the absence of Wnt signaling, β-catenin is found in complex in the cytoplasm with many proteins, including axin, PP2A, APC, GSK3, and CK1, which results in the phosphorylation of β-catenin by GSK-3β. Phosphorylated β-catenin is subsequently targeted for ubiquitination and proteosomal degradation. Axin plays a key role in assembling the destruction complex to reduce the β-catenin levels in the cytoplasm[32] as demonstrated by many studies.

Wnt signaling is initiated upon the binding of canonical Wnt ligands to their coreceptors Fz and LRP5/6 to stimulate the Wnt–β-catenin signaling pathway by destroying the β-catenin destruction complex (Fig. 3). Axin moves to the plasma membrane at the tail of LRP5/6 via interaction with dishevelled, which is bound on the tail of Fz. They dissociate the β-catenin destruction complex to form a complex which also includes CSk-3β and

FRAT1 (not shown) that prevents the phosphorylation of β-catenin by Csk-3β.[32,50] This leads to the accumulation of β-catenin in the cytoplasm and then its migration in the nucleus to activate osteoblast genes and thus promote bone formation.[83,84] Finally, Wnt can also regulate bone mass through a noncanonical pathway, which does not involve the utilization of LRP5/6 and activation of β-catenin.[85] Furthermore, Wnt can induce $[Ca^{2+}]i$ oscillations independently of the canonical and noncanonical pathways.[86] Hence, axin is an important regulator of Wnt signaling for osteoclast differentiation and bone formation.

5.3 RGS Proteins and Ca^{2+} Oscillations

Ca^{2+} is an important intracellular messenger that is critical for many cellular processes including cell differentiation.[40] Calcium levels can be maintained or regulated via cellular Ca^{2+} influxes and/or the stimulation-specific calcium channel. CaSR, a member of the GPCR family, acts as a Ca^{2+} sensor in the parathyroid and kidney to detect extracellular Ca^{2+} concentration and thus, helps control systemic calcium homeostasis by activating the $G\alpha_{q/11}$ pathway.[40] Activation of PLC then induces the formation of DAG and IP3, which then boost $[Ca^{2+}]i$ levels through activation of store-operated Ca^{2+} channel in the plasma membrane and Ca^{2+} release from the endoplasmic reticulum (ER). The effects of Ca^{2+} on cellular functions are mainly mediated through the Ca^{2+}-binding protein, CaM, which activates the phosphatase calcineurin,[87–89] which in turn dephosphorylates target genes for activation.[90,91] Conversely, an increase in intracellular Ca^{2+} levels can activate the plasma membrane Ca^{2+}ATPase and sarco/endoplasmic reticulum Ca^{2+} ATPase pumps to remove Ca^{2+} ions from the cytosol.[63]

While RGS proteins can regulate the duration of high Ca^{2+} states and control Ca^{2+} oscillations in different cell types by desensitizing the associated GPCRs through inactivation of $G\alpha_q$, the influences of Ca^{2+} oscillations on the regulation of many tissue functions and cell differentiation have not been reported.[92] RGS proteins, such as RGS1, RGS2, and RGS4, can bind to $G\alpha_q$ to enhance its GAP activity, leading to the inactivation of GPCR signaling and a decrease in $[Ca^{2+}]i$ in a cell-specific manner.[23] Consistent with that notion, deletion of RGS2 can reduce cell sensitivity to enhance Ca^{2+} oscillations from ER Ca^{2+} release into the cytosol.[63] Indeed, the amplitude and frequency of $[Ca^{2+}]i$ changes can influence cellular responses.[93,94] This is underscored by the findings that components of the Ca^{2+} signaling, including IP3 production and Ca^{2+} influxes across the ER and plasma membranes, can regulate the frequency and amplitude of Ca^{2+} oscillations.[16] Importantly, Ca^{2+} oscillation frequency may target cells along specific developmental pathways by differentially controlling the activation of distinct sets of genes.[95,96] Whereas rapid $[Ca^{2+}]i$ oscillations can stimulate the activation of NFAT, Oct/OAP, and NF-κB, infrequent $[Ca^{2+}]i$ oscillations can activate only NF-κB.

Similar to many cell types, Ca^{2+} is also critical for osteoblast and osteoclast differentiation.[40,67] CaSR is expressed in osteoblast precursors and can promote osteoblast differentiation in response to the elevated extracellular Ca^{2+} levels.[16] Also, recent reports have demonstrated that CaM kinase (CaMK) is expressed in osteoblasts, and pharmacological inhibition of CaMK can inhibit osteoblast differentiation both *in vivo* and *in vitro*.[40] Notably, PTH can influence Ca^{2+} oscillations and promote the expression of several genes in osteoblasts, including the matrix metallopeptidase 13 (Mmp13). Interestingly, CaMKII inhibition significantly decreases Mmp13 expression in osteoblasts in response to PTH stimulation, suggesting that some of the osteoblast responses to PTH are mediated by CaMKII. Moreover, another key osteoblastogenic agent, vitamin D3, can also activate CaMKII in response to Ca^{2+} influxes.[40] Collectively, these studies demonstrate that Ca^{2+} is a critical factor for osteoblast differentiation.

In osteoclast differentiation, RANKL can induce Ca^{2+} oscillations via the activation of PLC in osteoclast precursors through a crosstalk between RANK and the ITAM-based receptors Dap12 and FcRy, leading to the activation of calcineurin which dephosphorylates and activates NFATc1.[58,61] Hence, a threshold level of calcium is required to mediate a sustained level of NFATc1 for osteoclast differentiation.[61] Notably, the role of Ca^{2+} oscillations in osteoclast differentiation is buttressed by the reports showing that mice deficient in both Dap12 and FcRy display a severe osteopetrosis phenotype due to impaired Ca^{2+} oscillations and impaired NFATc1 activation.[58,60] Consistently, RBP-J has been identified as an inhibitor of osteoclastogenesis by its ability to interfere with RANKL-induced Ca^{2+} oscillations.[61] Moreover, pharmacological inhibition of CaM or calcineurin and silencing NFATc1, downstream of Ca^{2+} signaling, can all block osteoclast differentiation. Collectively, these studies support a critical for Ca^{2+} oscillations in osteoclastogenesis.

Given the critical influence of $[Ca^{2+}]i$ in osteoclast differentiation, we initially proposed that RGS proteins might play a key role in osteoclastogenesis by regulating Ca^{2+} oscillations. We found that RGS10 and RGS12 are specifically induced by RANKL during

Table 1 RGS Proteins in the Regulation of Osteoblasts and Osteoclasts

RGS Family Member	Gene Function in Bone Cells	Mechanism of Function in Bone Cells
RGS2	Promote osteoblast differentiation and function in bone formation	Regulation of PTH and Ca^{2+} signaling
Axin	Negatively regulate osteoblast differentiation and bone formation	Regulation of canonical Wnt signaling
RGS5	Promote osteoclast differentiation	Regulation of PTH levels
RGS10	Regulate osteoclast differentiation	Regulation of Ca^{2+} signaling and NFATc1 activation
RGS12	Regulate osteoclast differentiation	Regulation of Ca^{2+} oscillation via calcium channel and NFATc1 activation
RGS8	Inhibit osteoclast differentiation	Regulation of acid-sensing OGR1/NFATc1 signaling pathway

osteoclastogenesis.[66,68] RGS10 silencing inhibits osteoclastogenesis by blocking $[Ca^{2+}]$i levels *in vitro*, and deletion of RGS10 lead to a severe osteopetrosis.[66,67] Mechanistically, we revealed that RGS10 competitively interacts with Ca^{2+}/CaM and IP3 in a Ca^{2+}-dependent manner to activate PLC to evoke Ca^{2+} oscillations for NFATc1 activation and thereby promotes osteoclast differentiation. Similar to RGS10, we also found that RGS12 deficiency could also inhibit osteoclast differentiation by interfering with Ca^{2+} oscillations.[68] However, unlike RGS10, RGS12 does not bind to Ca^{2+}/CaM and IP3 in a Ca^{2+}-dependent manner during osteoclastogenesis, RGS12 can induce calcium influxes via activation of CaSR. These results support the notion that RGS proteins may regulate Ca^{2+} during osteoclast precursors in different ways to maintain a threshold level of Ca^{2+} which is required to osteoclastogenesis (Table 1).

6. GPCR/RGS SIGNALING IN SKELETAL DISORDERS

A variety of GPCRs, including Fz, GPR68, PTH1R, and CaSR, have been reported to play essential roles in skeletal development and/or bone homeostasis.[16,21] Hence, deletions or mutations of certain GPCRs have resulted in severe skeletal defects in mice. Also, many human diseases have been linked to defective GPCR signaling. As such, many drugs have been designed to target GPCRs for treating certain bone diseases. Nonetheless, whereas the roles of GPCRs in skeletal diseases have been highly studied, our understanding of RGS proteins in skeletal disorders is limited. Given the established roles of RGS proteins in regulating GPCR signaling, the mechanisms by which GPCRs cause skeletal diseases are likely to involve RGS proteins. Below, we provided some examples of skeletal diseases of defective GPCR signaling with emphasis into the aspects of signaling that are or may be regulated by RGS proteins.

6.1 PTH1R and GPCR 48 in Skeletal Development and Diseases

The roles of PTH and PTHrP, through activation of PTH1R, in bone are very complex. Whereas continuous administration of PTH or PTHrP results in bone loss from increased osteoclast activity, their intermittent administration triggers an increase in bone density from enhanced bone formation.[79,80] PTHR1 is expressed in bone and kidney and plays a critical role in Ca^{2+} homeostasis by activating the $G\alpha_q$–PLC signaling pathway. We have already discussed that Ca^{2+} is critical for both osteoblast and osteoclast differentiation. PTH1R can also activate the $G\alpha_s$–AC–cAMP signaling pathway. Notably, PTH and PTHrP can stimulate PTH1R on osteoblasts to increase the ratio of RANKL/OPG (osteoprotegrin), which in turn promotes osteoclastogenesis. OPG is a natural inhibitor of osteoclastogenesis which functions by competing with RANKL for RANK binding. Mice deficient in OPG display an osteoporosis phenotype from enhanced osteoclast formation and function, whereas transgenic mice with OPG overexpression have resulted in osteopetrosis from defective osteoclastogenesis.[97,98] The roles of PTH1R signaling in bone development and homeostasis have been established via numerous mice models in which this pathway is perturbed which results in severe skeletal defects.[21] Hence, the loss-of-function of PTH1R has been linked to Blomstrand osteochondrodyspasis type, a severe form of dwarfism, in humans.[99] Moreover, ligand-independent activation of PTH1R has been reported to cause Jansen's metaphyseal chondrodysplasia, a disease that is characterized mainly by short stature, normal ephiphyseal plates but disorganized metaphyseal regions, hypercalcemia, and hypophosphatemia. Also, defects in PTH1R can cause enchondromatosis and primary failure of tooth eruption in humans.[100]

Similar to PTH1R, but in a less complex manner, deletion of GPCR 48, a newly discovered glycoprotein hormone receptor subfamily of GPCRs, have been shown to affect skeletal development by inhibiting osteoblast differentiation while promoting osteoclast differentiation.[101] Interestingly, GPCR 48 can also activate the cAMP signaling pathway to reduce the expression level of Atf4 which leads to downregulation of Atf4 target genes in osteoblasts. Given that both PTH1R and GPCR48 activate the cAMP pathway which can be regulated by certain RGS proteins through attenuation of $G\alpha_s$–AC signaling, RGS proteins are likely to play a key role in regulating PTH1R and GPCR48 in both skeletal physiology and disease. This notion is underscored by a recent report demonstrating that generation of an engineered GPCR with constitutive active Gs in mice can induce a dramatic anabolic skeletal response by 9 weeks of age, which is associated with an increase in bone density, bone volume, and osteoblast markers.[102] So, the inability of the engineered GPCR to be inactivated by specific RGS proteins may be responsible for these bone effects under the constitutive activation of the Gs. Interestingly, this anabolic skeletal response was not observed during the first 4-week postnatal lives, suggesting that RGS protein may regulate bone formation in a temporal fashion. Further studies are needed to address this issue.

6.2 CaSR in Skeletal Development and Disease

As discussed above, CaSR can sense and respond to extracellular Ca^{2+} concentrations to regulate calcium homeostasis. Under elevated levels of plasma calcium, CaSR is activated, which in turn inhibit the release of PTH from the parathyroid gland. Moreover, CaSR can also function as a sensor for other cations such as lanthanum (La^{3+}), gadolinium (Gd^{3+}), barium (Ba^{2+}), strontium (Sr^{2+}), beryllium (Be^{2+}), and magnesium (Mg^{2+}).[103,104] Mice deficient in CaSR ($CaSR^{-/-}$) display severe rickets, characterized by defective hypertrophic chondrocytes, impaired growth plate calcification, disordered mineral deposition, excessive osteoid accumulation, and prolonged mineralization lag time in metaphyseal bone.[105] Also, the $CaSR^{-/-}$ mice developed hyperparathyroidism, hypercalcemia, and dwarfism as well as increased bone mineralization and resorption. To elucidate the mechanism of CaSR functions in bone, Dvorak and colleagues generated a transgenic mouse with constitutively active CaSR in mature osteoblasts driven by the osteocalcin promoter.[106] These mice show an increase in RANKL expression by osteoblasts which trigger an increase in osteoclast formation and bone resorption. These transgenic mice also show an increase in bone formation. Consistently, CaSR can upregulate PTHrP in osteoblasts to promote osteoblast and osteoclast differentiation in part via an increase in extracellular Ca^{2+} levels and RANKL upregulation.[16]

In line with the mouse studies, inactivating mutations of the CaSR gene have been reported to cause many diseases in humans, including the familial hypocalciuric hypercalcemia (FHH) and neonatal severe hyperparathyroidism (NSHPT).[107,108] NSHPT can cause a life-threatening hypercalcemia and defectively mineralized skeleton, whereas FHH can show mild-to-moderate hypercalcemia. Moreover, mutations that activate CaSR can cause autosomal dominant hypocalcemia (ADH) in humans. ADH patients display low-to-normal levels of PTH and hypercalciuria with hypocalcemia but no overt skeletal deformities. Ligand binding to CaSR can activate the PLC signaling pathway in both osteoblasts and osteoclasts, presumably via the activation of $G\alpha_q$, to increase intracellular calcium levels. A role for RGS proteins in regulating the functions of CaSR in the skeletal system was demonstrated by our reports that RGS10 and RGS12 can cause skeletal defects by interfering with Ca^{2+} oscillations via the PLC pathways and Ca^{2+} influxes through the calcium channel, respectively.[66–68] More studies are needed to explicitly address the roles of other RGS in regulating Ca^{2+} signaling in skeletal physiology and bone disorders.

6.3 The Canonical Wnt Signaling Pathway in Bone Disease

Wnt signaling pathway is a critical regulator of many development processes, including skeletal development.[32] In addition, Wnt signaling has also been associated with many bone diseases. Wnt signaling is a very complicated pathway that includes the canonical Wnt–β-catenin, the noncanonical Wnt-planar cell polarity, and the Wnt–calcium (Wnt–Ca^{2+}) pathways. Of the three Wnt signaling components, however, the canonical Wnt–Ca^{2+} pathway has been shown to be the major component of the Wnt signaling that regulates bone cells. So, activation of Wnt signaling prevents β-catenin degradation through obliteration of the multiprotein β-catenin destruction complex and thus increases the β-catenin level in the cytoplasm, from where it is translocated in the nucleus to induce gene expression for cell differentiation and function (Fig. 3).

6.3.1 Axin in Bone Development and Bone Disease

Axin is a key component of the β-catenin destruction complex.[109,110] The role of axin in suppressing the canonical Wnt–β-catenin signaling pathway has been addressed above (Fig. 4). Consistent with these findings, targeted deletion of axin in mice causes abnormal skull development, a phenotype resembling craniosynostosis in humans, resulting from an increase in calvarial osteoblast development.[50–52] These mutant mice also display defective bone ossification from enhanced osteoblast proliferation and differentiation. The skeletal anomalies of the axin-deficient mice were shown to be mediated through enhanced activation of β-catenin. This finding is in agreement with that of Yan *et al.* who demonstrated that axin also functions as a negative regulator of bone remodeling in adult mice and promotes osteoblastogenesis through the β-catenin–BMP2/4–Osx signaling pathway.[50] Taken together, these findings demonstrate a crucial role for this RGS protein in regulating the canonical Wnt–β-catenin signaling pathway in skeletal development and bone disease. Given the central of axin in the regulating Wnt signaling, we further examined the role of Wnt in skeletal development and bone disease.

6.3.2 Mouse Models of Wnt Signaling in Skeletal Homeostasis

The role of the canonical Wnt–β-catenin in bone development and homeostasis is underscored by the findings from many mouse models in which this signaling pathway is disturbed which lead to skeletal deformities.[32] Whereas mutations that enhance the Wnt–β-catenin signaling pathway in osteoblasts can trigger an increase in bone mineralization, its inactivation in the osteoblastic lineage cells can result in a decrease in bone mass. For instance, alteration of the Wnt10b gene, one type of Wnt ligand, has established it as a positive regulator of bone homeostasis by enhancing osteoblast differentiation in adult bone.[111–113] Conversely, deletion of Wls, a chaperone required for the secretion of Wnt proteins, causes severe osteoporosis in adult mice through impairment of osteoblast differentiation and enhancement of osteoclast differentiation.[114] Furthermore, mice with mutations in the LRP5 and/or LPR6 have been shown to recapitulate human diseases of defective Wnt signaling. Hence, global LRP5 deficiency or its targeted deletion in osteocytes has recapitulated the osteoporotic low-bone formation phenotype of humans with loss-of-function mutations in LRP5 or LRP6.[115–117] Conversely, introduction of the human high-bone mass gain-of-function mutations in LRP5 β-propeller 1 has induced a similar phenotype in mice when expressed ubiquitously or selectively in the limbs or in the cells of the osteoblastic lineage.[32] Further studies have identified many endogenous enhancers of Wnt signaling, such as four R-spondin (RSPO) proteins which bind to Fz and possibly LRP6 to induce β-catenin activation and endogenous inhibitors of Wnt signaling such as sclerostin and dickkopf (DKK1) which can bind to LRP5 or LRP6 to inhibit Wnt signaling.[32] The Wnt signaling in bone is complex, and our discussion here only highlights some aspects of the canonical Wnt–β-catenin signaling pathway and its relevance to bone disease.

6.3.3 Bone Diseases of Defective Wnt Signaling

Many human diseases have been reported to be associated with abnormal Wnt signaling leading to skeletal deformities.[32] Some of these diseases result from LRP5 mutations. Loss-of-function mutations in LPR5 can lead to osteoporosis-pseudoglioma syndrome, an autosomal recessive disorder characterized by severe juvenile-onset osteoporosis from decreased bone formation and congenital or juvenile-onset blindness, and other diseases with similar phenotypes.[118–120] Gain-of-function mutations in LRP5 can cause high bone-mass syndrome from increased bone formation. Notably, gain-of-function mutations of the LRP5 gene can decrease LPR5 binding with sclerostin and DKK1, which in turn promote the activation of the Wnt–β-catenin signaling pathway. Other mutations have shown to affect the Wnt signaling by targeting sclerostin.[121,122] Indeed, sclerostin deficiency can cause sclerosteosis and Van Buchen disease, two autosomal recessive disorders characterized by bone overgrowth from excessive bone formation mostly in skull and mandible.[123] These human reports provide strong evidence that the Wnt–β-catenin signaling pathway is critical for skeletal homeostasis and bone development. Also, inactivating mutations that affect the Wnt signaling cause severe skeletal deformities.

6.4 RGS Proteins in Inflammatory Bone Disease

Excessive osteoclast formation and/or activity are responsible for the tooth loss associated with many oral diseases including periodontitis. Many RGS proteins including RGS10 can promote osteoclasts differentiation, and proinflammatory factors, such as tumor necrosis factor alpha (TNFα), interleukin 1 (IL-1), and IL-6, can potently influence osteoclast differentiation and function.[124–126] In fact, the trio of RANKL, TNFα, and IL-1 are known to be strongly associated with bone loss in many oral diseases.[126–129] We reasoned that given its role in osteoclasts, RGS10 might also play a role in boss induced by inflammatory conditions like periodontitis. Toward this end, we employed the adeno-associated virus (AAV)-mediated gene silencing strategy to deplete RGS10 (AAV-shRNA-RGS10) in an established mouse model of bacteria-induced periodontal disease. Our data revealed that RGS10 silencing can protect mice from both the inflammation and bone loss associated with the periodontal disease. We showed that the number of dendritic cells, T cells, and osteoclasts were all decreased in the disease mice-treated with AAV-shRNA-RGS10 as compared to control littermates. Notably, the expression of osteoclast markers and proinflammatory markers were also reduced in the periodontal lesions. These results indicate that RGS10 can regulate both osteoclast differentiation and inflammatory bone diseases.

Our finding is in agreement with that of a previous report which showed that human monocyte-derived dendritic cells constitutively express high levels of RGS2, RGS10, RGS14, RGS18, and RGS19, but low levels of RGS3 and RGS13.[38] Activation of toll-like receptor 3 (TLR3) or TLR4 on these dendritic cells can stimulate the expression of RGS1, RGS16, and RGS20 and attenuate the expression levels of RGS18 and RGS14 without modifying other RGS proteins. These results indicate that TLR signaling can affect GPCR signaling by altering the expression of RGS proteins. The role of RGS in macrophages was addressed by Lee and colleagues who demonstrated that macrophages deficient in RGS10 ($RGS10^{-/-}$) generate higher levels of TNF, IL-1β, and IL-12p70 in response to LPS treatment and also exhibited higher cytotoxicity on dopaminergic MN9D neuroblastoma cells.[130] Moreover, the authors found that $Rgs10^{-/-}$ macrophages display dysregulated M1 responses along with blunted M2 alternative activation responses. While our finding on the role of RGS10 in periodontitis may seem in disagreement with that of this study, these discrepancies may stem from the complex nature of RGS protein functions which can regulate GPCR signaling in a temporal- and cell-specific manner. Nonetheless, more studies are needed to fully investigate the role RGS10 in inflammatory conditions. We anticipate that RGS10, aside from its reported role in periodontitis, may also carry critical roles in other inflammatory bone diseases like rheumatoid arthritis.

7. CONCLUSION AND PERSPECTIVES

Our understanding of the roles of RGS proteins in the skeletal system remains mostly unexplored aside from a few studies that have demonstrated essential roles for some RGS proteins in bone formation and/or homeostasis as well as bone diseases through their regulatory effects on osteoblasts, chondrocytes, and/or osteoclasts. Whereas these limited numbers of studies have helped improve our understanding of GPCR–G proteins–RGS signaling in bone biology, much work remains to be done in this area. Besides, GPCR signaling controls many physiological processes, so impacting the functions of RGS proteins in the skeletal system may undesirably have impacts on other organ systems or vice versa. Consequently, it will be critical to address this issue as we further our understanding of RGS proteins on bone biology with the hope of developing better therapy for bone loss by specifically and efficiently targeting GPCR signaling. Indeed, GPCR signaling has been and is likely to remain a valuable target for many disease states, including allergy, ulcers and reflux, pain, high blood pressure, migraine headache, nausea, schizophrenia, and depression. Notably, therapy targeting sclerostin is being examined for treating osteoporosis.[32] Most importantly, through their regulatory roles on GPCR signaling and their cell-specific and -selectivity patterns of functions, targeting RGS proteins may also evolve as an effective and potent therapeutic option for bone loss associated with bone disorders of either excessive osteoclast function or attenuated osteoblast activity.

ACKNOWLEDGMENTS

Research reported in this publication was supported by the National Institute of Athritis and Musculoskeletal and Skin Diseases, part of the National Institutes of Health, under AR-055307 and R01-AR-44741 to Y.P.L.

REFERENCES

1. Feng X. Chemical and biochemical basis of cell-bone matrix interaction in health and disease. *Curr Chem Biol.* 2009;3(2):189–196.
2. Boyce BF, Rosenberg E, de Papp AE, Duong le T. The osteoclast, bone remodelling and treatment of metabolic bone disease. *Eur J Clin Invest.* 2012;42(12):1332–1341.
3. Asagiri M, Takayanagi H. The molecular understanding of osteoclast differentiation. *Bone.* 2007;40(2):251–264.
4. Karsenty G. Transcriptional control of skeletogenesis. *Annu Rev Genomics Hum Genet.* 2008;9:183–196.
5. Karsenty G, Oury F. Biology without walls: the novel endocrinology of bone. *Annu Rev Physiol.* 2012;74:87–105.
6. Provot S, Schipani E. Molecular mechanisms of endochondral bone development. *Biochem Biophys Res Commun.* 2005;328(3):658–665.
7. Teitelbaum SL. Bone resorption by osteoclasts. *Science.* 2000;289(5484):1504–1508.
8. Zhao Q, Shao J, Chen W, Li YP. Osteoclast differentiation and gene regulation. *Front Biosci.* 2007;12:2519–2529.
9. Boyle WJ, Simonet WS, Lacey DL. Osteoclast differentiation and activation. *Nature.* 2003;423(6937):337–342.
10. Feng X, McDonald JM. Disorders of bone remodeling. *Annu Rev Pathol.* 2011;6:121–145.
11. Jules J, Ashley JW, Feng X. Selective targeting of RANK signaling pathways as new therapeutic strategies for osteoporosis. *Expert Opin Ther Targets.* 2010;14(9):923–934.
12. Page RC. Milestones in periodontal research and the remaining critical issues. *J Periodontal Res.* 1999;34(7):331–339.
13. Goronzy JJ, Weyand CM. Rheumatoid arthritis. *Immunol Rev.* 2005;204:55–73.

14. Lee DM, Weinblatt ME. Rheumatoid arthritis. *Lancet*. 2001;358(9285):903–911.
15. Kimple AJ, Bosch DE, Giguere PM, Siderovski DP. Regulators of G-protein signaling and their Galpha substrates: promises and challenges in their use as drug discovery targets. *Pharmacol Rev*. 2011;63(3):728–749.
16. Keinan D, Yang S, Cohen RE, Yuan X, Liu T, Li YP. Role of regulator of G protein signaling proteins in bone. *Front Biosci*. 2014;19:634–648.
17. McCudden CR, Hains MD, Kimple RJ, Siderovski DP, Willard FS. G-protein signaling: back to the future. *Cell Mol Life Sci*. 2005;62(5):551–577.
18. Bansal G, Druey KM, Xie Z. R4 RGS proteins: regulation of G-protein signaling and beyond. *Pharmacol Ther*. 2007;116(3):473–495.
19. Zalewska M, Siara M, Sajewicz W. G protein-coupled receptors: abnormalities in signal transmission, disease states and pharmacotherapy. *Acta Pol Pharm*. 2014;71(2): 229–243.
20. Louwette S, Van Geet C, Freson K. Regulators of G protein signaling: role in hematopoiesis, megakaryopoiesis and platelet function. *J Thromb Haemost*. 2012;10(11):2215–2222.
21. Wu M, Deng L, Zhu G, Li YP. G protein and its signaling pathway in bone development and disease. *Front Biosci*. 2010;15:957–985.
22. Morris AJ, Malbon CC. Physiological regulation of G protein-linked signaling. *Physiol Rev*. 1999;79(4):1373–1430.
23. Ross EM, Wilkie TM. GTPase-activating proteins for heterotrimeric G proteins: regulators of G protein signaling (RGS) and RGS-like proteins. *Annu Rev Biochem*. 2000;69:795–827.
24. Doupnik CA, Davidson N, Lester HA, Kofuji P. RGS proteins reconstitute the rapid gating kinetics of gbetagamma-activated inwardly rectifying K+ channels. *Proc Natl Acad Sci U S A*. 1997;94(19):10461–10466.
25. De Vries L, Zheng B, Fischer T, Elenko E, Farquhar MG. The regulator of G protein signaling family. *Annu Rev Pharmacol Toxicol*. 2000;40:235–271.
26. Hollinger S, Hepler JR. Cellular regulation of RGS proteins: modulators and integrators of G protein signaling. *Pharmacol Rev*. 2002;54(3):527–559.
27. Xie GX, Palmer PP. How regulators of G protein signaling achieve selective regulation. *J Mol Biol*. 2007;366(2):349–365.
28. Terzi D, Stergiou E, King SL, Zachariou V. Regulators of G protein signaling in neuropsychiatric disorders. *Prog Mol Biol Transl Sci*. 2009;86:299–333.
29. Traynor J. mu-Opioid receptors and regulators of G protein signaling (RGS) proteins: from a symposium on new concepts in mu-opioid pharmacology. *Drug Alcohol Depend*. 2012;121(3):173–180.
30. Wieland T, Lutz S, Chidiac P. Regulators of G protein signalling: a spotlight on emerging functions in the cardiovascular system. *Curr Opin Pharmacol*. 2007;7(2):201–207.
31. Hurst JH, Hooks SB. Regulator of G-protein signaling (RGS) proteins in cancer biology. *Biochem Pharmacol*. 2009;78(10):1289–1297.
32. Baron R, Kneissel M. WNT signaling in bone homeostasis and disease: from human mutations to treatments. *Nat Med*. 2013;19(2):179–192.
33. Miles RR, Sluka JP, Santerre RF, et al. Dynamic regulation of RGS2 in bone: potential new insights into parathyroid hormone signaling mechanisms. *Endocrinology*. 2000;141(1):28–36.
34. Thirunavukkarasu K, Halladay DL, Miles RR, Geringer CD, Onyia JE. Analysis of regulator of G-protein signaling-2 (RGS-2) expression and function in osteoblastic cells. *J Cell Biochem*. 2002;85(4):837–850.
35. Eszlinger M, Holzapfel HP, Voigt C, Arkenau C, Paschke R. RGS 2 expression is regulated by TSH and inhibits TSH receptor signaling. *Eur J Endocrinol*. 2004;151(3):383–390.

36. Riekenberg S, Farhat K, Debarry J, et al. Regulators of G-protein signalling are modulated by bacterial lipopeptides and lipopolysaccharide. *FEBS J.* 2009;276(3):649–659.
37. Lee HK, Yeo S, Kim JS, et al. Protein kinase C-eta and phospholipase D2 pathway regulates foam cell formation via regulator of G protein signaling 2. *Mol Pharmacol.* 2010;78(3):478–485.
38. Shi GX, Harrison K, Han SB, Moratz C, Kehrl JH. Toll-like receptor signaling alters the expression of regulator of G protein signaling proteins in dendritic cells: implications for G protein-coupled receptor signaling. *J Immunol.* 2004;172(9):5175–5184.
39. Giorelli M, Livrea P, Defazio G, et al. Interferon beta-1a counteracts effects of activation on the expression of G-protein-coupled receptor kinases 2 and 3, beta-arrestin-1, and regulators of G-protein signalling 2 and 16 in human mononuclear leukocytes. *Cell Signal.* 2002;14(8):673–678.
40. Zayzafoon M. Calcium/calmodulin signaling controls osteoblast growth and differentiation. *J Cell Biochem.* 2006;97(1):56–70.
41. Klein RF, Nissenson RA, Strewler GJ. Pertussis toxin inhibits hormonal stimulation of bone resorption in fetal rat limb bones. *J Pharmacol Exp Ther.* 1991;258(3):877–881.
42. Ko JK, Choi KH, Kim IS, Jung EK, Park DH. Inducible RGS2 is a cross-talk regulator for parathyroid hormone signaling in rat osteoblast-like UMR106 cells. *Biochem Biophys Res Commun.* 2001;287(4):1025–1033.
43. Abou-Samra AB, Juppner H, Force T, et al. Expression cloning of a common receptor for parathyroid hormone and parathyroid hormone-related peptide from rat osteoblast-like cells: a single receptor stimulates intracellular accumulation of both cAMP and inositol trisphosphates and increases intracellular free calcium. *Proc Natl Acad Sci U S A.* 1992;89(7):2732–2736.
44. Silve CM, Hradek GT, Jones AL, Arnaud CD. Parathyroid hormone receptor in intact embryonic chicken bone: characterization and cellular localization. *J Cell Biol.* 1982;94(2):379–386.
45. Homme M, Schaefer F, Mehls O, Schmitt CP. Differential regulation of RGS-2 by constant and oscillating PTH concentrations. *Calcif Tissue Int.* 2009;84(4):305–312.
46. Koh J, Dar M, Untch BR, et al. Regulator of G protein signaling 5 is highly expressed in parathyroid tumors and inhibits signaling by the calcium-sensing receptor. *Mol Endocrinol.* 2011;25(5):867–876.
47. Miao D, Tong XK, Chan GK, Panda D, McPherson PS, Goltzman D. Parathyroid hormone-related peptide stimulates osteogenic cell proliferation through protein kinase C activation of the Ras/mitogen-activated protein kinase signaling pathway. *J Biol Chem.* 2001;276(34):32204–32213.
48. Brown EM. Mechanisms underlying the regulation of parathyroid hormone secretion in vivo and in vitro. *Curr Opin Nephrol Hypertens.* 1993;2(4):541–551.
49. Liang Y, Li C, Guzman VM, et al. Identification of a novel alternative splicing variant of RGS5 mRNA in human ocular tissues. *FEBS J.* 2005;272(3):791–799.
50. Yan Y, Tang D, Chen M, et al. Axin2 controls bone remodeling through the beta-catenin-BMP signaling pathway in adult mice. *J Cell Sci.* 2009;122(pt 19):3566–3578.
51. Gwak J, Hwang SG, Park HS, et al. Small molecule-based disruption of the axin/beta-catenin protein complex regulates mesenchymal stem cell differentiation. *Cell Res.* 2012;22(1):237–247.
52. Minear S, Leucht P, Jiang J, et al. Wnt proteins promote bone regeneration. *Sci Transl Med.* 2010;2(29):29ra30.
53. Yu HM, Jerchow B, Sheu TJ, et al. The role of Axin2 in calvarial morphogenesis and craniosynostosis. *Development.* 2005;132(8):1995–2005.
54. Hay E, Laplantine E, Geoffroy V, et al. N-cadherin interacts with axin and LRP5 to negatively regulate Wnt/beta-catenin signaling, osteoblast function, and bone formation. *Mol Cell Biol.* 2009;29(4):953–964.

55. Regard JB, Cherman N, Palmer D, et al. Wnt/beta-catenin signaling is differentially regulated by Galpha proteins and contributes to fibrous dysplasia. *Proc Natl Acad Sci U S A*. 2011;108(50):20101–20106.
56. Takayanagi H, Kim S, Koga T, et al. Induction and activation of the transcription factor NFATc1 (NFAT2) integrate RANKL signaling in terminal differentiation of osteoclasts. *Dev Cell*. 2002;3(6):889–901.
57. Crabtree GR, Olson EN. NFAT signaling: choreographing the social lives of cells. *Cell*. 2002;109(suppl):S67–S79.
58. Koga T, Inui M, Inoue K, et al. Costimulatory signals mediated by the ITAM motif cooperate with RANKL for bone homeostasis. *Nature*. 2004;428(6984):758–763.
59. Zou W, Kitaura H, Reeve J, et al. Syk, c-Src, the alphavbeta3 integrin, and ITAM immunoreceptors, in concert, regulate osteoclastic bone resorption. *J Cell Biol*. 2007;176(6):877–888.
60. Mocsai A, Humphrey MB, Van Ziffle JA, et al. The immunomodulatory adapter proteins DAP12 and Fc receptor gamma-chain (FcRgamma) regulate development of functional osteoclasts through the Syk tyrosine kinase. *Proc Natl Acad Sci U S A*. 2004;101(16):6158–6163.
61. Li S, Miller CH, Giannopoulou E, Hu X, Ivashkiv LB, Zhao B. RBP-J imposes a requirement for ITAM-mediated costimulation of osteoclastogenesis. *J Clin Invest*. 2014;124(11):5057–5073.
62. Zhao B, Grimes SN, Li S, Hu X, Ivashkiv LB. TNF-induced osteoclastogenesis and inflammatory bone resorption are inhibited by transcription factor RBP-J. *J Exp Med*. 2012;209(2):319–334.
63. Wang X, Huang G, Luo X, Penninger JM, Muallem S. Role of regulator of G protein signaling 2 (RGS2) in Ca(2+) oscillations and adaptation of Ca(2+) signaling to reduce excitability of $RGS2^{-/-}$ cells. *J Biol Chem*. 2004;279(40):41642–41649.
64. Brown EM. Role of the calcium-sensing receptor in extracellular calcium homeostasis. *Best Pract Res Clin Endocrinol Metab*. 2013;27(3):333–343.
65. Lombardi G, Di Somma C, Rubino M, et al. The roles of parathyroid hormone in bone remodeling: prospects for novel therapeutics. *J Endocrinol Invest*. 2011;34(7 suppl):18–22.
66. Yang S, Chen W, Stashenko P, Li YP. Specificity of RGS10A as a key component in the RANKL signaling mechanism for osteoclast differentiation. *J Cell Sci*. 2007;120(pt 19):3362–3371.
67. Yang S, Li YP. RGS10-null mutation impairs osteoclast differentiation resulting from the loss of [Ca2+]i oscillation regulation. *Genes Dev*. 2007;21(14):1803–1816.
68. Yang S, Li YP. RGS12 is essential for RANKL-evoked signaling for terminal differentiation of osteoclasts in vitro. *J Bone Miner Res*. 2007;22(1):45–54.
69. Schiff ML, Siderovski DP, Jordan JD, et al. Tyrosine-kinase-dependent recruitment of RGS12 to the N-type calcium channel. *Nature*. 2000;408(6813):723–727.
70. Richman RW, Diverse-Pierluissi MA. Mapping of RGS12-Cav2.2 channel interaction. *Methods Enzymol*. 2004;390:224–239.
71. Richman RW, Strock J, Hains MD, et al. RGS12 interacts with the SNARE-binding region of the Cav2.2 calcium channel. *J Biol Chem*. 2005;280(2):1521–1528.
72. Iwai K, Koike M, Ohshima S, et al. RGS18 acts as a negative regulator of osteoclastogenesis by modulating the acid-sensing OGR1/NFAT signaling pathway. *J Bone Miner Res*. 2007;22(10):1612–1620.
73. Tomura H, Mogi C, Sato K, Okajima F. Proton-sensing and lysolipid-sensitive G-protein-coupled receptors: a novel type of multi-functional receptors. *Cell Signal*. 2005;17(12):1466–1476.
74. Ludwig MG, Vanek M, Guerini D, et al. Proton-sensing G-protein-coupled receptors. *Nature*. 2003;425(6953):93–98.

75. Huang C, Hydo LM, Liu S, Miller RT. Activation of choline kinase by extracellular Ca2+ is Ca(2+)-sensing receptor, Galpha12 and Rho-dependent in breast cancer cells. *Cell Signal*. 2009;21(12):1894–1900.
76. Heino TJ, Chagin AS, Savendahl L. The novel estrogen receptor G-protein-coupled receptor 30 is expressed in human bone. *J Endocrinol*. 2008;197(2):R1–R6.
77. Uzawa T, Hori M, Ejiri S, Ozawa H. Comparison of the effects of intermittent and continuous administration of human parathyroid hormone(1–34) on rat bone. *Bone*. 1995;16(4):477–484.
78. Tam CS, Heersche JN, Murray TM, Parsons JA. Parathyroid hormone stimulates the bone apposition rate independently of its resorptive action: differential effects of intermittent and continuous administration. *Endocrinology*. 1982;110(2):506–512.
79. Takeuchi Y. Treatment of osteoporosis with PTH. *Clin Calcium*. 2014;24(6):893–902.
80. Inoue D. Role for PTHrP in bone and cartilage metabolism. *Clin Calcium*. 2014;24(6):863–869.
81. Ueno Y, Shinki T, Nagai Y, Murayama H, Fujii K, Suda T. In vivo administration of 1,25-dihydroxyvitamin D3 suppresses the expression of RANKL mRNA in bone of thyroparathyroidectomized rats constantly infused with PTH. *J Cell Biochem*. 2003;90(2):267–277.
82. Homme M, Schmitt CP, Himmele R, Hoffmann GF, Mehls O, Schaefer F. Vitamin D and dexamethasone inversely regulate parathyroid hormone-induced regulator of G protein signaling-2 expression in osteoblast-like cells. *Endocrinology*. 2003;144(6):2496–2504.
83. Cadigan KM, Peifer M. Wnt signaling from development to disease: insights from model systems. *Cold Spring Harb Perspect Biol*. 2009;1(2):a002881.
84. Kim W, Kim M, Jho EH. Wnt/beta-catenin signalling: from plasma membrane to nucleus. *Biochem J*. 2013;450(1):9–21.
85. Jenny A. Planar cell polarity signaling in the Drosophila eye. *Curr Top Dev Biol*. 2010;93:189–227.
86. Kohn AD, Moon RT. Wnt and calcium signaling: beta-catenin-independent pathways. *Cell Calcium*. 2005;38(3–4):439–446.
87. Hogan PG, Chen L, Nardone J, Rao A. Transcriptional regulation by calcium, calcineurin, and NFAT. *Genes Dev*. 2003;17(18):2205–2232.
88. Im SH, Rao A. Activation and deactivation of gene expression by Ca^{2+}/calcineurin-NFAT-mediated signaling. *Mol Cells*. 2004;18(1):1–9.
89. Rao A, Luo C, Hogan PG. Transcription factors of the NFAT family: regulation and function. *Annu Rev Immunol*. 1997;15:707–747.
90. Brown EM, Gamba G, Riccardi D, et al. Cloning and characterization of an extracellular Ca(2+)-sensing receptor from bovine parathyroid. *Nature*. 1993;366(6455):575–580.
91. Garrett JE, Capuano IV, Hammerland LG, et al. Molecular cloning and functional expression of human parathyroid calcium receptor cDNAs. *J Biol Chem*. 1995;270(21):12919–12925.
92. Luo X, Popov S, Bera AK, Wilkie TM, Muallem S. RGS proteins provide biochemical control of agonist-evoked [Ca2+]i oscillations. *Mol Cell*. 2001;7(3):651–660.
93. Dolmetsch RE, Lewis RS. Signaling between intracellular Ca2+ stores and depletion-activated Ca2+ channels generates [Ca2+]i oscillations in T lymphocytes. *J Gen Physiol*. 1994;103(3):365–388.
94. Dolmetsch RE, Xu K, Lewis RS. Calcium oscillations increase the efficiency and specificity of gene expression. *Nature*. 1998;392(6679):933–936.
95. Fields RD, Eshete F, Stevens B, Itoh K. Action potential-dependent regulation of gene expression: temporal specificity in ca2+, cAMP-responsive element binding proteins, and mitogen-activated protein kinase signaling. *J Neurosci*. 1997;17(19):7252–7266.

96. Gu X, Spitzer NC. Distinct aspects of neuronal differentiation encoded by frequency of spontaneous Ca^{2+} transients. *Nature*. 1995;375(6534):784–787.
97. Simonet WS, Lacey DL, Dunstan CR, et al. Osteoprotegerin: a novel secreted protein involved in the regulation of bone density. *Cell*. 1997;89(2):309–319.
98. Wu Y, Liu J, Guo H, et al. Establishment of OPG transgenic mice and the effect of OPG on bone microarchitecture. *Int J Endocrinol*. 2013;2013:125932.
99. Smith EP, Boyd J, Frank GR, et al. Estrogen resistance caused by a mutation in the estrogen-receptor gene in a man. *N Engl J Med*. 1994;331(16):1056–1061.
100. Decker E, Stellzig-Eisenhauer A, Fiebig BS, et al. PTHR1 loss-of-function mutations in familial, nonsyndromic primary failure of tooth eruption. *Am J Hum Genet*. 2008;83(6):781–786.
101. Luo J, Zhou W, Zhou X, et al. Regulation of bone formation and remodeling by G-protein-coupled receptor 48. *Development*. 2009;136(16):2747–2756.
102. Hsiao EC, Boudignon BM, Chang WC, et al. Osteoblast expression of an engineered Gs-coupled receptor dramatically increases bone mass. *Proc Natl Acad Sci U S A*. 2008;105(4):1209–1214.
103. Theman TA, Collins MT. The role of the calcium-sensing receptor in bone biology and pathophysiology. *Curr Pharm Biotechnol*. 2009;10(3):289–301.
104. Chattopadhyay N, Brown EM. Cellular "sensing" of extracellular calcium (Ca(2+)(o)): emerging roles in regulating diverse physiological functions. *Cell Signal*. 2000;12(6):361–366.
105. Garner SC, Pi M, Tu Q, Quarles LD. Rickets in cation-sensing receptor-deficient mice: an unexpected skeletal phenotype. *Endocrinology*. 2001;142(9):3996–4005.
106. Dvorak MM, Chen TH, Orwoll B, et al. Constitutive activity of the osteoblast Ca^{2+}-sensing receptor promotes loss of cancellous bone. *Endocrinology*. 2007;148(7):3156–3163.
107. Brown EM, Pollak M, Chou YH, Seidman CE, Seidman JG, Hebert SC. Cloning and functional characterization of extracellular Ca(2+)-sensing receptors from parathyroid and kidney. *Bone*. 1995;17(2 suppl):7S–11S.
108. Pollak MR, Brown EM, Chou YH, et al. Mutations in the human Ca(2+)-sensing receptor gene cause familial hypocalciuric hypercalcemia and neonatal severe hyperparathyroidism. *Cell*. 1993;75(7):1297–1303.
109. Jho EH, Zhang T, Domon C, Joo CK, Freund JN, Costantini F. Wnt/beta-catenin/Tcf signaling induces the transcription of Axin2, a negative regulator of the signaling pathway. *Mol Cell Biol*. 2002;22(4):1172–1183.
110. Gerlach JP, Emmink BL, Nojima H, Kranenburg O, Maurice MM. Wnt signalling induces accumulation of phosphorylated beta-catenin in two distinct cytosolic complexes. *Open Biol*. 2014;4(11):140120.
111. Bennett CN, Ouyang H, Ma YL, et al. Wnt10b increases postnatal bone formation by enhancing osteoblast differentiation. *J Bone Miner Res*. 2007;22(12):1924–1932.
112. Bennett CN, Longo KA, Wright WS, et al. Regulation of osteoblastogenesis and bone mass by Wnt10b. *Proc Natl Acad Sci U S A*. 2005;102(9):3324–3329.
113. Stevens JR, Miranda-Carboni GA, Singer MA, Brugger SM, Lyons KM, Lane TF. Wnt10b deficiency results in age-dependent loss of bone mass and progressive reduction of mesenchymal progenitor cells. *J Bone Miner Res*. 2010;25(10):2138–2147.
114. Zhong Z, Zylstra-Diegel CR, Schumacher CA, et al. Wntless functions in mature osteoblasts to regulate bone mass. *Proc Natl Acad Sci U S A*. 2012;109(33):E2197–E2204.
115. Kato M, Patel MS, Levasseur R, et al. Cbfa1-independent decrease in osteoblast proliferation, osteopenia, and persistent embryonic eye vascularization in mice deficient in Lrp5, a Wnt coreceptor. *J Cell Biol*. 2002;157(2):303–314.

116. Clement-Lacroix P, Ai M, Morvan F, et al. Lrp5-independent activation of Wnt signaling by lithium chloride increases bone formation and bone mass in mice. *Proc Natl Acad Sci U S A.* 2005;102(48):17406–17411.
117. Cui Y, Niziolek PJ, MacDonald BT, et al. Lrp5 functions in bone to regulate bone mass. *Nat Med.* 2011;17(6):684–691.
118. Gong Y, Slee RB, Fukai N, et al. LDL receptor-related protein 5 (LRP5) affects bone accrual and eye development. *Cell.* 2001;107(4):513–523.
119. Little RD, Carulli JP, Del Mastro RG, et al. A mutation in the LDL receptor-related protein 5 gene results in the autosomal dominant high-bone-mass trait. *Am J Hum Genet.* 2002;70(1):11–19.
120. Boyden LM, Mao J, Belsky J, et al. High bone density due to a mutation in LDL-receptor-related protein 5. *N Engl J Med.* 2002;346(20):1513–1521.
121. Poole KE, van Bezooijen RL, Loveridge N, et al. Sclerostin is a delayed secreted product of osteocytes that inhibits bone formation. *FASEB J.* 2005;19(13):1842–1844.
122. Brunkow ME, Gardner JC, Van Ness J, et al. Bone dysplasia sclerosteosis results from loss of the SOST gene product, a novel cystine knot-containing protein. *Am J Hum Genet.* 2001;68(3):577–589.
123. Balemans W, Patel N, Ebeling M, et al. Identification of a 52 kb deletion downstream of the SOST gene in patients with van Buchem disease. *J Med Genet.* 2002;39(2):91–97.
124. Jules J, Shi Z, Liu J, Xu D, Wang S, Feng X. Receptor activator of NF-{kappa}B (RANK) cytoplasmic IVVY535-538 motif plays an essential role in tumor necrosis factor-{alpha} (TNF)-mediated osteoclastogenesis. *J Biol Chem.* 2010;285(48): 37427–37435.
125. Jules J, Zhang P, Ashley JW, et al. Molecular basis of requirement of receptor activator of nuclear factor kappaB signaling for interleukin 1-mediated osteoclastogenesis. *J Biol Chem.* 2012;287(19):15728–15738.
126. Kikuta J, Yamaguchi M, Shimizu M, Yoshino T, Kasai K. Notch signaling induces root resorption via RANKL and IL-6 from hPDL cells. *J Dent Res.* 2014;94:140–147.
127. Akiyama T, Miyamoto Y, Yoshimura K, et al. Porphyromonas gingivalis-derived lysine gingipain enhances osteoclast differentiation induced by tumor necrosis factor-alpha and interleukin-1beta but suppresses that by interleukin-17A: importance of proteolytic degradation of osteoprotegerin by lysine gingipain. *J Biol Chem.* 2014;289(22):15621–15630.
128. Meng S, Zhang L, Tang Y, et al. BET inhibitor JQ1 blocks inflammation and bone destruction. *J Dent Res.* 2014;93(7):657–662.
129. Taubman MA, Valverde P, Han X, Kawai T. Immune response: the key to bone resorption in periodontal disease. *J Periodontol.* 2005;76(11 suppl):2033–2041.
130. Lee JK, Chung J, Kannarkat GT, Tansey MG. Critical role of regulator G-protein signaling 10 (RGS10) in modulating macrophage M1/M2 activation. *PLoS One.* 2013;8(11):e81785.

CHAPTER FIVE

Regulator of G Protein Signaling 2: A Versatile Regulator of Vascular Function

Patrick Osei-Owusu*[,1], Kendall J. Blumer[†]

*Department of Pharmacology and Physiology, Drexel University College of Medicine, Philadelphia, Pennsylvania, USA

[†]Department of Cell Biology and Physiology, Washington University School of Medicine, St. Louis, Missouri, USA

[1]Corresponding author: e-mail address: patrick.osei-owusu@drexelmed.edu

Contents

Abstract

Regulators of G protein signaling (RGS) proteins of the B/R4 family are widely expressed in the cardiovascular system where their role in fine-tuning G protein signaling is critical to maintaining homeostasis. Among members of this family, RGS2 and RGS5 have been shown to play key roles in cardiac and smooth muscle function by tightly regulating signaling pathways that are activated through $G_{q/11}$ and $G_{i/o}$ classes of heterotrimeric G proteins. This chapter reviews accumulating evidence supporting a key role for RGS2 in vascular function and the implication of changes in RGS2 function and/or expression in the pathogenesis of blood pressure disorders, particularly hypertension. With such understanding, RGS2 and the signaling pathways it controls may emerge as novel targets for developing next-generation antihypertensive drugs/agents.

Progress in Molecular Biology and Translational Science, Volume 133
ISSN 1877-1173
http://dx.doi.org/10.1016/bs.pmbts.2015.02.001

1. INTRODUCTION

G protein-coupled receptor (GPCR) signaling mediates the biological effects of many physiologically important vasoactive substances such as norepinephrine, epinephrine, angiotensin II (ANG II), endothelin-1 (ET-1), vasopressin, acetylcholine (ACh), and bradykinin. These vasoactive substances activate GPCRs that couple to one or more of four heterotrimeric G protein families distinguished by the type of α subunit in the heterotrimer (G_s, $G_{q/11}$, $G_{i/o}$, and $G_{12/13}$).[1–4] RGS proteins play key roles in G protein signaling by acting in part as GTPase-activating proteins (GAPs) that accelerate the rate that Gα subunits hydrolyze GTP and consequently deactivate signal transduction. By facilitating reformation of inactive GDP-bound $G_{\alpha\beta\gamma}$ heterotrimers, this mechanism also resets the system for further rounds of activation by agonist-activated GPCRs. Growing evidence indicates that RGS proteins also regulate cell signaling by mechanisms independent of GAP activity.[5–7]

In vertebrates, RGS proteins are expressed in essentially all cell types, tissues, and organ systems where they play critical roles in physiology and disease.[8,9] Among >30 RGS proteins encoded by the human or mouse genome, RGS2 has proved to have diverse functions in the cardiovascular system, including blood pressure regulation. Whereas studies of RGS2 and related B/R4-class RGS proteins in cardiovascular biology and other processes were reviewed several years ago,[10] significant progress has been made since then.[11–13] Accordingly, this chapter reviews current understanding of RGS2 structure, function, and regulation in vascular biology, and highlights current questions that are driving research in the field.

2. STRUCTURE AND BIOCHEMICAL FUNCTIONS OF RGS2

In mice and humans, the *Rgs2* locus is located on chromosome 1 and contains five exons. This locus encodes a 212-residue protein containing a RGS domain of ~120 amino acids that is flanked by an ~80-residue N-terminal domain and a short C-terminal tail,[14] similar to other B/R4-class RGS proteins.[15] RGS2 possesses intrinsic GAP activity that is potent and selective *in vitro* for G_q-class Gα subunits,[16,17] in contrast to other B/R4-family RGS proteins that have intrinsic GAP activity for both G_q- and $G_{i/o}$-class Gα subunits. However, GAP activity of RGS2 toward $G_{i/o}$-class Gα subunits can be detected in receptor-driven systems,[18] suggesting that interaction with GPCRs or other components influence which G proteins can be regulated by RGS2. RGS2 possesses an N-terminal amphipathic α-helix and a hydrophobic motif that facilitate targeting of the protein to the plasma membrane where GPCRs and G proteins reside.[14] Plasma membrane targeting of RGS2 is enhanced by cGMP-dependent protein kinase Iα-mediated phosphorylation of serine 46 and 64.[19] The N-terminal domain of RGS2 also binds other proteins, including spinophilin, muscarinic and α-adrenergic receptors, and directly inhibits certain adenylyl cyclase isoforms.[6,20–24] Furthermore, RGS2 is capable of regulating protein synthesis independently of its GAP activity by binding elongation factor 2b.[5]

3. REGULATION OF RGS2 EXPRESSION

RGS2 expression is regulated at the levels of transcription and protein degradation. RGS2 was first identified as an immediate early gene in activated blood mononuclear cells.[25] Subsequent investigations in many cell types identified various stimuli and signaling pathways that induce *Rgs2* gene transcription. For example, ANG II- and G_s-coupled GPCR agonists and pharmacological agents that increase cAMP levels or intracellular Ca^{2+} increase RGS2 mRNA expression in several cell types, including neonatal cardiomyocytes, neurons, HEK293, and vascular smooth muscle cell lines.[26–28] Moreover, CREB elements in the *Rgs2* gene promoter have been shown to drive expression in response to GPCR agonists.[29]

Like many regulatory proteins whose transcription is tightly controlled, the half-life of RGS2 protein is short (e.g., ~30 min in vascular smooth muscle cells[19]) due to proteasome-mediated degradation. Degradation of RGS2 is augmented by a polymorphism (RGS2-Q2L) identified initially in a cohort of Japanese hypertension patients[30,31] and subsequently in other genetic studies of human hypertension.[32,33] Defining how RGS2 is degraded may provide avenues to increase its expression level and function in diseases linked to RGS2 deficiency or dysfunction, including hypertension, anxiety disorders, and airway hyperreactivity.[33–36]

4. EXPRESSION OF RGS2 IN THE CARDIOVASCULAR SYSTEM

RGS2 is widely expressed in the cardiovascular system including brain, heart, vasculature, and kidney.[37] Whereas all B/R4-class RGS proteins are expressed at the mRNA level in myocardium,[38] only a few (RGS1, 2, 4, and 5) have been shown thus far to be expressed detectably at the mRNA or protein level in vascular smooth muscle or endothelial cells of the arterial system.[39–43] In mouse heart, RGS2 has been found to regulate pathologic remodeling elicited by pressure overload.[44,45] In this model, RGS2 deficiency increases the susceptibility and extent of pathologic cardiac hypertrophic responses that progress rapidly to dilated heart failure. Cardiomyocyte-specific overexpression of $G_q\alpha$ in RGS2 knockout mice is sufficient to trigger cardiac hypertrophy and dilated cardiomyopathy, indicating that impaired regulation of G_q signaling by RGS2 in cardiomyocytes plays a causal role.[44] Moreover, loss of G_q-dependent regulation by RGS2 enhances ANG II-induced profibrotic responses, which can be reversed by adenovirus-mediated overexpression of RGS2 in ventricular myocytes.[45]

5. RGS2 FUNCTION IN THE VASCULATURE

5.1 Smooth Muscle Function

Homozygous and heterozygous RGS2 knockout mice exhibit elevated blood pressure,[40] indicating that the mouse *Rgs2* gene is a quantitative trait locus affecting blood pressure. Consistent with the function of RGS2 as a G_q GAP, agonists that activate G_q-coupled GPCRs evoke augmented Ca^{2+} transients in vascular smooth muscle cells from mesenteric resistance arteries of RGS2 knockout mice.[19] Furthermore, RGS2 deficiency enhances contraction of mesenteric resistance arteries or renal interlobar arteries stimulated by an α_1-adrenergic receptor agonist.[46,47] Thus, G_q-mediated smooth muscle contraction is augmented by RGS2 deficiency in several regions of the arterial tree (Fig. 1).

RGS2 also promotes smooth muscle intrinsic relaxation by functioning as an effector of the nitric oxide/cGMP-dependent protein kinase Iα (PKGIα) signaling pathway that attenuates G_q-evoked vasoconstriction. In this mechanism, nitric oxide (NO) produced by endothelial cells stimulates production of cGMP in vascular smooth muscle cells to activate PKGIα. RGS2 is phosphorylated by PKGIα on serine-46 and -64, which promotes its association with the plasma membrane and attenuation of G_q-evoked vasoconstriction.[19,48] Indeed, the absence of RGS2 abolishes the ability of cGMP analogs to blunt vasopressin-evoked Ca^{2+} transients in resistance artery smooth muscle cells.[49] Moreover, RGS2 deficiency blunts the ability of a NO donor to antagonize the pressor effect of an

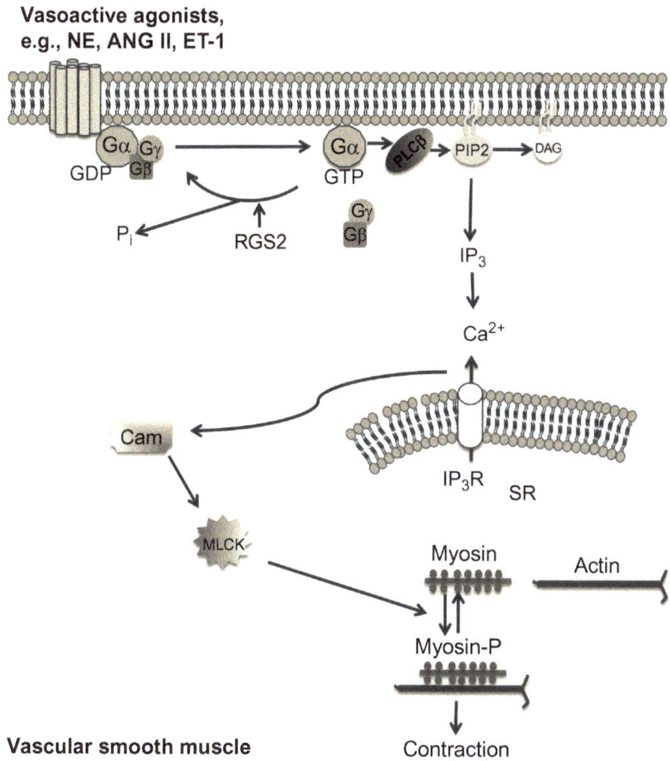

Figure 1 The role of RGS2 in vascular smooth muscle contraction. RGS2 tightly regulates the duration and amplitude of G protein signaling that mediates smooth muscle contraction triggered by vasoactive ligands that activate G_q-coupled GPCRs. Upon activation, G_q signaling evokes a rise in intracellular Ca^{2+} concentration resulting in myosin phosphorylation by myosin light chain kinase (MLCK) and actomyosin crossbridge formation. Decreased RGS2 expression or function can lead to increased signaling duration and intensity that in turn cause augmented smooth muscle contraction. Cam, calmodulin; DAG, diacylglycerol; PIP_2, phosphatidylinositol 4,5-bisphosphate; PLCβ, phospholipase C-beta; IP_3, inositol 1,3,5-trisphosphate; SR, sarcoplasmic reticulum; IP_3R, IP_3 receptor; NE, norepinephrine; ET-1, endothelin-1; ANG II, angiotensin II. (See the color plate.)

α_1-adrenergic receptor agonist that directly stimulates contraction of the resistance vasculature.[49]

RGS2 deficiency also appears to augment vasoconstriction of resistance arteries by affecting stretch-induced contraction or myogenic tone. Renal interlobar arteries from RGS2 knockout mice exhibit enhanced constriction triggered by rises in transmural pressure *ex vivo*.[46] How RGS2 regulates myogenic tone is unknown. Myogenic response evoked by stretch of smooth muscle cells depends on activation of stretch receptors mediating mechanotransduction.[50,51] Although the components of this signaling mechanism in smooth muscle remain to be fully defined, receptors studied to date include members of the transient receptor potential superfamily,[52] vascular epithelial sodium channel,[53] integrins,[54] and the angiotensin type 1 receptor (AT1R).[55] Whereas RGS2 is known to negatively regulate vascular AT1R signaling, it remains possible that RGS2 regulates other mediators of vascular mechanotransduction (Fig. 2) that are not known to couple to heterotrimeric G proteins. Because mechanotransduction occurs prominently in resistance as compared to conduit arteries, elucidating how RGS2 regulates mechanotransduction may reveal new insight into mechanisms of vascular dysfunction in hypertension or other cardiovascular diseases in which augmented vascular resistance occurs.

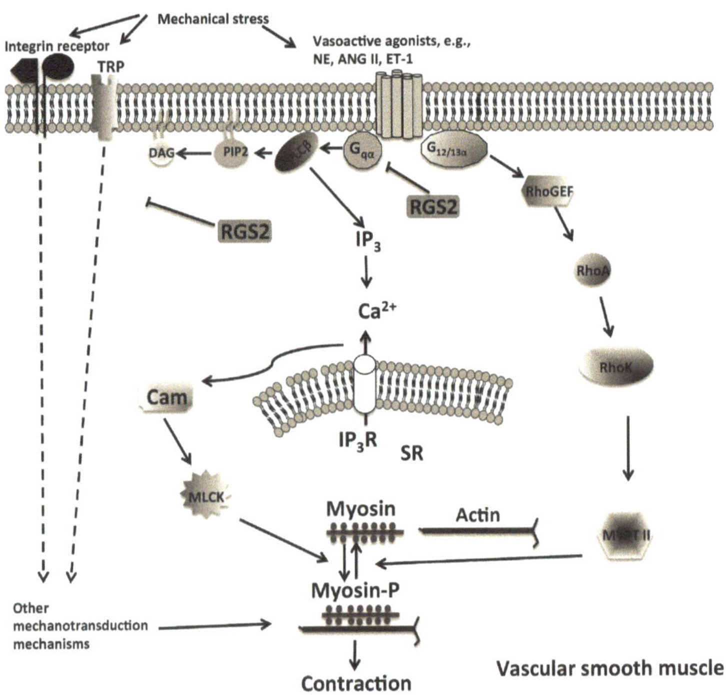

Figure 2 A model of contractile signaling pathways potentially regulated by RGS2 in smooth muscle cells in the resistance vasculature. In addition to regulating G_q-dependent smooth muscle contraction, RGS2 could potentially regulate stretch-induced contraction mediated by putative receptors and ion channels that can be activated by mechanical stimuli. TRP, transient receptor potential; NE, norepinephrine; ET-1, endothelin-1; ANG II, angiotensin II; myosin-P, phosphorylated myosin; RhoK, RhoA-associated kinase; MYPT II, myosin phosphatase regulatory subunit II. (See the color plate.)

5.2 Endothelial Function

Because RGS2 is expressed in vascular endothelium, smooth muscle, and adventitia,[56,57] it potentially regulates vascular function in several ways. Indeed, recent studies have revealed a novel endothelium-specific function for RGS2.[47] Because G_q signaling in vascular endothelium elicits Ca^{2+} fluxes to stimulate production of NO and endothelium-derived hyperpolarizing factor (EDHF), RGS2 deficiency would be expected to augment G_q activity and consequent endothelium-dependent vasodilatation. Surprisingly, however, the opposite effect has been found. Global or endothelium-specific knockout of RGS2 markedly impairs endothelium-dependent relaxation of mesenteric resistance arteries stimulated by ACh.[47] Whereas global RGS2 deficiency causes relatively modest impairment of ACh-evoked relaxation mediated by NO, it nearly completely eliminates ACh-evoked relaxation mediated by EDHF. This defect in EDHF-mediated relaxation can be rescued by blocking $G_{i/o}$ signaling with pertussis toxin. Thus, in contrast to its role in vascular smooth muscle as a GAP for G_q, RGS2 apparently functions in endothelium of mesenteric resistance arteries as a GAP that squelches $G_{i/o}$ signaling, which otherwise would inhibit EDHF-dependent vasodilatation (Fig. 3).

In contrast to its effect on endothelial function in mesenteric resistance arteries, RGS2 deficiency does not significantly affect endothelium-dependent relaxation of aortic rings.[48] This finding is consistent with evidence indicating that the relative contributions of various endothelium-derived relaxing factors, including NO, EDHF, and prostacyclin, change as the diameter of the vessel lumen decreases from conduit vessels where NO predominates, to small arterioles, where EDHF is a primary mediator of endothelium-dependent vasodilation.[58,59] However, it is inconsistent with evidence indicating that RGS2 deficiency significantly impairs relaxation of aortic rings evoked by cGMP, which is a direct effector of the NO system. Despite such unresolved questions, what remains clear is that RGS2 has distinct functions in different types of arteries and in different compartments within arteries.

Although significant understanding of RGS2 function in the vasculature has been obtained, an important unanswered question is how RGS2 deficiency causes hypertension. Whereas global RGS2 deficiency causes hypertension, endothelium- or vascular smooth muscle-specific deletion of RGS2 is insufficient to increase blood pressure.[47] However, transplantation of kidneys lacking RGS2 into wild-type recipient mice is sufficient to

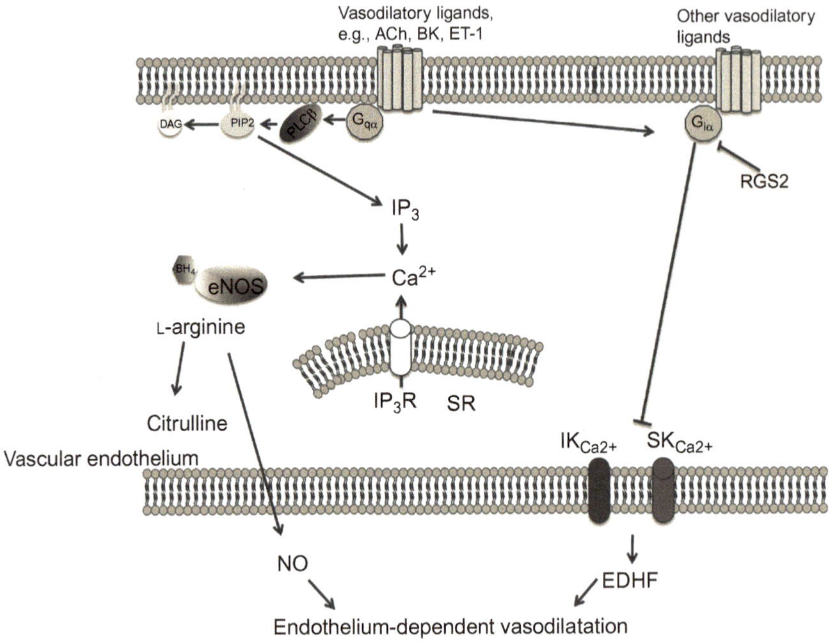

Figure 3 Regulation of G protein signaling in vascular endothelium. Vasodilatory ligands such as ACh and BK activate G_q-coupled receptors in the vascular endothelium of resistance arteries causing a rise in intracellular Ca^{2+}, required to stimulate endothelial nitric oxide synthase (eNOS) for producing nitric oxide (NO). A rise in endothelial Ca^{2+} is also necessary for activation of potassium channels SK_{Ca} and IK_{Ca} that mediate the production of endothelium-derived hyperpolarizing factor (EDHF). The activity of SK_{Ca} and IK_{Ca} are negatively regulated by Gα subunits of the G_i-class in this vascular compartment. By acting as GAP, RGS2 squelches the activity of G_i-class G proteins to promote EDHF production. Thus, RGS2 deficiency in the endothelium is proposed to result in impairment of endothelium-dependent vasodilatation due to increased G_i signaling, inhibiting EDHF production. IP_3, inositol 1,4,5-trisphosphate; PLCβ, phospholipase C-beta; BH_4, tetrahydrobiopterin; eNOS, endothelium nitric oxide synthase. (See the color plate.)

elevate blood pressure,[60] as expected, since the kidney plays a dominant role in regulating blood pressure. Because RGS2 is expressed widely throughout the kidney, including renal vasculature and parenchyma,[46,61] future experiments that analyze the effects of deleting RGS2 in specific renal structures and cell types will be required to assess the vascular and tubular contributions of this RGS protein in renal control of blood pressure homeostasis. Such studies may reveal whether integrated versus compartment-specific functions of RGS2 in renal vasculature and tubules are required for normal regulation of blood pressure.

6. RGS2 AND HUMAN HYPERTENSION

The initial discovery of hypertension in heterozygous or homozygous, RGS2 knockout mice[40] prompted investigations whether RGS2 dysfunction occurs in human hypertension. Initial studies examined RGS2 expression and function in human skin fibroblasts and peripheral blood mononuclear cells (PBMNC) in normotensive and hypertensive individuals.[62] They found that cells from hypertensive subjects exhibited decreased RGS2 mRNA expression and increased ANG II-induced ERK phosphorylation and Ca^{2+} release from internal stores.[62] Human genetic studies subsequently have identified *Rgs2* single nucleotide polymorphisms (SNPs) in hypertension cohorts from several regions of the globe including, Japan,[31] the Netherlands,[32] United States,[33] and China.[63,64] Some of these SNPs apparently decrease RGS2 protein expression by increasing its susceptibility to arginylation/ubiquitination and subsequent degradation by the proteasome pathway.[30] Other mutations or SNPs impair GAP activity or affect the subcellular localization of RGS2.[36] Whether or how these SNPs affect RGS2-dependent control of organ systems critical for blood pressure homeostasis remains to be determined.

In contrast to evidence linking decreased expression or function of RGS2 in hypertension, evidence of augmented RGS2 expression has been reported in Bartter/Gitelman syndrome, a human disorder characterized by hypotension, and sodium and potassium wasting due to defects in renal electrolyte transport proteins.[65,66] Interestingly, PBMNC from Bartter/Gitelman syndrome patients show elevated RGS2 expression, and refractory response to ANG II-induced Ca^{2+} mobilization from internal stores.[57] Thus, identifying regulatory mechanisms that up- or downregulate RGS2 expression or function to increase or decrease blood pressure may provide new targets for drug discovery in cardiovascular diseases.

7. RGS2 IN PREECLAMPSIA

Pregnancy triggers physiological remodeling of the cardiovascular system that is necessary to maintain normal blood pressure and organ perfusion in the face of increased maternal extracellular fluid volume.[67,68] Marked increase in the levels of vasoactive hormones, particularly ANG II, occur during pregnancy, which has the potential to augment vasoconstriction, peripheral resistance, sodium retention, and blood pressure.[69–71] However, such effects ordinarily are countered by increased production of endothelium-derived relaxing factors, mainly NO.[72] Furthermore, the arterial vasculature in pregnancy becomes refractory to vasoconstriction by ANG II,[73] although the mechanisms involved remain poorly understood. Defects in such compensatory mechanisms, therefore, may contribute to gestational hypertension and preeclampsia.[74,75]

RGS2 may function in compensatory mechanisms that normally maintain blood pressure during pregnancy and are impaired in preeclampsia. This hypothesis is attractive because ANG II triggers vasoconstriction by activating G_q-coupled AT_1 receptors such that increased expression or activity of RGS2 or other G_q regulatory proteins could provide adaptive mechanisms that normally maintain blood pressure during pregnancy. Impairment of such adaptive mechanisms potentially could contribute to the pathophysiology of preeclampsia. Indeed, recent studies of G protein signaling pathways associated with blood pressure regulation in preeclamptic and normotensive pregnant women identified the *Rgs2* SNP rs4606 in the 3′ untranslated region as associated with risk and progression of preeclampsia.[76] Further evidence indicated that women experiencing preeclampsia and carrying this *Rgs*2 SNP have increased risk of developing hypertension after delivery.[77] These findings should motivate further studies aimed at unraveling how G protein signaling is functionally remodeled during pregnancy, whether RGS2 is an important component of this regulatory network, and whether RGS2 may provide a therapeutic target in preeclampsia.

8. RGS2 IN OTHER DISORDERS AFFECTING SMOOTH MUSCLE

Because contraction and relaxation of smooth muscle in a variety of organs and tissues occur by similar mechanisms, the functions of RGS2 established in the vasculature may be recapitulated in other organ systems and diseases. This hypothesis has been explored in asthma and chronic obstructive pulmonary disease (COPD), a central feature of which is increased constriction and spasms of airway smooth muscle driven by various GPCRs, principally G_q-coupled muscarinic receptors.[78] Recent studies have shown that RGS2 expression is decreased in lungs of asthmatic relative to nonasthmatic individuals,[35] and that RGS2 knockout mice exhibit spontaneous hyperresponsiveness of airway smooth muscle accompanied by augmented GPCR-induced intracellular Ca^{2+} mobilization.[79] In humans, the *Rgs*2 SNPs rs2746071 and rs2746072 are in linkage disequilibrium, exhibiting higher prevalence in asthmatics compared to controls.[79]

Moreover, RGS2 expression in human airway smooth muscle is increased by currently used therapeutics for asthma and COPD, including long-acting β_2-adrenergic agonists alone or combined with glucocorticoids.[80] Thus, RGS2 dysfunction in some patients may play a causal role in airway smooth muscle hypercontraction, and upregulation of RGS2 may improve therapeutic outcome in asthma or COPD patients.

9. CONCLUSIONS AND FUTURE PERSPECTIVES

RGS2 has emerged as a critical component of G protein signaling mechanisms in various systems that regulate cardiovascular function. In the heart, RGS2 tightly regulates G_q signaling to maintain normal cardiac structure and function.[44] Similarly, G_q regulation by RGS2 in the vasculature is key to maintaining normal vascular tone by controlling constriction triggered by ligands that activate G_q-coupled GPCRs.[40] In addition, by serving as a substrate of the NO–cGMP–PKG signaling pathway, RGS2 promotes smooth muscle intrinsic relaxation, thereby reducing vascular tone.[19,48] Moreover, RGS2 in endothelium reduces vascular tone by acting as a GAP toward $G_{i/o}$-class Gα proteins to facilitate EDHF-dependent vasodilatation of resistance arteries.[47] These roles of RGS2 are consistent with studies showing that defective RGS2 regulation of G_q- and $G_{i/o}$-class Gα proteins play a primary role in the pathogenesis of cardiovascular and other smooth muscle disorders such as heart failure, COPD, asthma, and hypertension. Thus, mechanisms that impinge on RGS2 expression could be critical to disease onset or progression.

Certain *Rgs2* SNPs or mutations identified to date that are linked to human diseases are capable of altering RGS2 expression and/or function.[36] Following the discovery more than a decade ago that RGS2 has an important role in blood pressure homeostasis,[40] subsequent human genetics studies have identified association of several *Rgs*2 SNPs with hypertension in various populations.[32,33] Some of the hypertension-linked SNPs of *Rgs2* have been shown to decrease RGS2 GAP activity toward G_q *in vitro*.[36] In contrast, decreased RGS2 expression is associated with increased G_q signaling and hypotension in Bartter/Gitelman syndrome.[57] These studies highlight the relevance of the function of RGS2 in human cardiovascular function.

Several outstanding questions must be answered to obtain a more complete understanding of how RGS2 regulates vascular function in normal physiology and disease. First, how does RGS2 deficiency cause hypertension? Recent transplantation studies establish that RGS2 deficiency in the kidney is sufficient to increase blood pressure,[60] although defining the renal mechanisms that are dysregulated in the absence of RGS2 remains an important goal. Second, what mechanisms functionally shift the GAP activity of RGS2 toward $G_{i/o}$- as opposed to G_q-class Gα proteins in the vascular endothelium? Addressing this question may have significant impact because endothelial dysfunction is a central feature of several cardiovascular disorders, particularly hypertension and preeclampsia, which are also associated with *Rgs2* SNPs.[33,76] Third, what mechanisms regulate RGS2 expression levels, and how are they altered to up- or downregulate RGS2 in health and disease? Unraveling these mechanisms may provide novel therapeutic targets for increasing or decreasing the endogenous level of RGS2 in diseases and disorders as diverse as hypertension, asthma, COPD, Bartter's/ Gitelman's syndrome, and anxiety that have been linked to RGS2.

ACKNOWLEDGMENTS

We would like to thank Dr. Harpreet Singh for critical reading of the manuscript. The authors were supported by NIH grants HL075632 and GM44592 to K.J.B., and institutional support from Drexel University College of Medicine to P.O.

REFERENCES

1. Wess J. G-protein-coupled receptors: molecular mechanisms involved in receptor activation and selectivity of G-protein recognition. *FASEB J.* 1997;11:346–354.
2. Hur EM, Kim KT. G protein-coupled receptor signalling and cross-talk: achieving rapidity and specificity. *Cell Signal.* 2002;14:397–405.
3. Conklin BR, Bourne HR. Structural elements of G alpha subunits that interact with G beta gamma, receptors, and effectors. *Cell.* 1993;73:631–641.
4. Neer EJ. Heterotrimeric G proteins: organizers of transmembrane signals. *Cell.* 1995;80:249–257.
5. Nguyen CH, Ming H, Zhao P, et al. Translational control by RGS2. *J Cell Biol.* 2009;186:755–765. http://dx.doi.org/10.1083/jcb.200811058.
6. Roy AA, Baragli A, Bernstein LS, Hepler JR, Hebert TE, Chidiac P. RGS2 interacts with Gs and adenylyl cyclase in living cells. *Cell Signal.* 2006;18:336–348. http://dx.doi.org/10.1016/j.cellsig.2005.05.004.
7. Sethakorn N, Yau DM, Dulin NO. Non-canonical functions of RGS proteins. *Cell Signal.* 2010;22:1274–1281. http://dx.doi.org/10.1016/j.cellsig.2010.03.016.
8. Hollinger S, Hepler JR. Cellular regulation of RGS proteins: modulators and integrators of G protein signaling. *Pharmacol Rev.* 2002;54:527–559.
9. Zhang P, Mende U. Functional role, mechanisms of regulation, and therapeutic potential of regulator of G protein signaling 2 in the heart. *Trends Cardiovasc Med.* 2014;24:85–93. http://dx.doi.org/10.1016/j.tcm.2013.07.002.
10. Bansal G, Druey KM, Xie Z. R4 RGS proteins: regulation of G-protein signaling and beyond. *Pharmacol Ther.* 2007;116:473–495. http://dx.doi.org/10.1016/j.pharmthera.2007.09.005.
11. Nisancioglu MH, Mahoney Jr WM, Kimmel DD, Schwartz SM, Betsholtz C, Genove G. Generation and characterization of rgs5 mutant mice. *Mol Cell Biol.* 2008;28:2324–2331. http://dx.doi.org/10.1128/MCB.01252-07.

12. Cho H, Park C, Hwang IY, et al. Rgs5 targeting leads to chronic low blood pressure and a lean body habitus. *Mol Cell Biol.* 2008;28:2590–2597. http://dx.doi.org/10.1128/MCB.01889-07.
13. Grillet N, Pattyn A, Contet C, Kieffer BL, Goridis C, Brunet JF. Generation and characterization of Rgs4 mutant mice. *Mol Cell Biol.* 2005;25:4221–4228. http://dx.doi.org/10.1128/MCB.25.10.4221-4228.2005.
14. Heximer SP, Lim H, Bernard JL, Blumer KJ. Mechanisms governing subcellular localization and function of human RGS2. *J Biol Chem.* 2001;276:14195–14203. http://dx.doi.org/10.1074/jbc.M009942200.
15. Siderovski DP, Willard FS. The GAPs, GEFs, and GDIs of heterotrimeric G-protein alpha subunits. *Int J Biol Sci.* 2005;1:51–66.
16. Heximer SP, Watson N, Linder ME, Blumer KJ, Hepler JR. RGS2/G0S8 is a selective inhibitor of Gqalpha function. *Proc Natl Acad Sci USA.* 1997;94:14389–14393.
17. Nance MR, Kreutz B, Tesmer VM, Sterne-Marr R, Kozasa T, Tesmer JJ. Structural and functional analysis of the regulator of G protein signaling 2-galphaq complex. *Structure.* 2013;21:438–448. http://dx.doi.org/10.1016/j.str.2012.12.016.
18. Ingi T, Krumins AM, Chidiac P, et al. Dynamic regulation of RGS2 suggests a novel mechanism in G-protein signaling and neuronal plasticity. *J Neurosci.* 1998;18:7178–7188.
19. Osei-Owusu P, Sun X, Drenan RM, Steinberg TH, Blumer KJ. Regulation of RGS2 and second messenger signaling in vascular smooth muscle cells by cGMP-dependent protein kinase. *J Biol Chem.* 2007;282:31656–31665. http://dx.doi.org/10.1074/jbc.M706360200.
20. Salim S, Dessauer CW. Analysis of the interaction between RGS2 and adenylyl cyclase. *Methods Enzymol.* 2004;390:83–99. http://dx.doi.org/10.1016/S0076-6879(04)90006-7.
21. Salim S, Sinnarajah S, Kehrl JH, Dessauer CW. Identification of RGS2 and type V adenylyl cyclase interaction sites. *J Biol Chem.* 2003;278:15842–15849. http://dx.doi.org/10.1074/jbc.M210663200.
22. Wang X, Zeng W, Soyombo AA, et al. Spinophilin regulates Ca^{2+} signalling by binding the N-terminal domain of RGS2 and the third intracellular loop of G-protein-coupled receptors. *Nat Cell Biol.* 2005;7:405–411. http://dx.doi.org/10.1038/ncb1237.
23. Bernstein LS, Ramineni S, Hague C, et al. RGS2 binds directly and selectively to the M1 muscarinic acetylcholine receptor third intracellular loop to modulate Gq/11alpha signaling. *J Biol Chem.* 2004;279:21248–21256. http://dx.doi.org/10.1074/jbc.M312407200.
24. Hague C, Bernstein LS, Ramineni S, Chen Z, Minneman KP, Hepler JR. Selective inhibition of alpha1A-adrenergic receptor signaling by RGS2 association with the receptor third intracellular loop. *J Biol Chem.* 2005;280:27289–27295. http://dx.doi.org/10.1074/jbc.M502365200.
25. Siderovski DP, Heximer SP, Forsdyke DR. A human gene encoding a putative basic helix-loop-helix phosphoprotein whose mRNA increases rapidly in cycloheximide-treated blood mononuclear cells. *DNA Cell Biol.* 1994;13:125–147.
26. Zou MX, Roy AA, Zhao Q, Kirshenbaum LA, Karmazyn M, Chidiac P. RGS2 is upregulated by and attenuates the hypertrophic effect of alpha1-adrenergic activation in cultured ventricular myocytes. *Cell Signal.* 2006;18:1655–1663. http://dx.doi.org/10.1016/j.cellsig.2006.01.012.
27. Grant SL, Lassegue B, Griendling KK, Ushio-Fukai M, Lyons PR, Alexander RW. Specific regulation of RGS2 messenger RNA by angiotensin II in cultured vascular smooth muscle cells. *Mol Pharmacol.* 2000;57:460–467.
28. Song L, De Sarno P, Jope RS. Muscarinic receptor stimulation increases regulators of G-protein signaling 2 mRNA levels through a protein kinase C-dependent mechanism. *J Biol Chem.* 1999;274:29689–29693.

29. Xie Z, Liu D, Liu S, et al. Identification of a cAMP-response element in the regulator of G-protein signaling-2 (RGS2) promoter as a key cis-regulatory element for RGS2 transcriptional regulation by angiotensin II in cultured vascular smooth muscles. *J Biol Chem.* 2011;286:44646–44658. http://dx.doi.org/10.1074/jbc.M111.265462.
30. Bodenstein J, Sunahara RK, Neubig RR. N-terminal residues control proteasomal degradation of RGS2, RGS4, and RGS5 in human embryonic kidney 293 cells. *Mol Pharmacol.* 2007;71:1040–1050. http://dx.doi.org/10.1124/mol.106.029397.
31. Yang J, Kamide K, Kokubo Y, et al. Genetic variations of regulator of G-protein signaling 2 in hypertensive patients and in the general population. *J Hypertens.* 2005;23:1497–1505.
32. Hahntow IN, Mairuhu G, van Valkengoed IG, et al. Are RGS2 gene polymorphisms associated with high blood pressure in an ethnicity- and gender-specific manner? *Am J Hypertens.* 2009;22:80–86. http://dx.doi.org/10.1038/ajh.2008.310.
33. Riddle EL, Rana BK, Murthy KK, et al. Polymorphisms and haplotypes of the regulator of G protein signaling-2 gene in normotensives and hypertensives. *Hypertension.* 2006;47:415–420. http://dx.doi.org/10.1161/01.HYP.0000200714.81990.61.
34. Hettema JM, Sun C, Chen X, Kendler KS. Genetic association study between RGS2 and anxiety-related phenotypes. *Psychiatr Genet.* 2013;23:92. http://dx.doi.org/10.1097/YPG.0b013e32835d70b3.
35. Xie Y, Jiang H, Nguyen H, et al. Regulator of G protein signaling 2 is a key modulator of airway hyperresponsiveness. *J Allergy Clin Immunol.* 2012;130. http://dx.doi.org/10.1016/j.jaci.2012.05.004, 968.e3–976.e3.
36. Gu S, Tirgari S, Heximer SP. The RGS2 gene product from a candidate hypertension allele shows decreased plasma membrane association and inhibition of Gq. *Mol Pharmacol.* 2008;73:1037–1043. http://dx.doi.org/10.1124/mol.107.044214.
37. Kehrl JH, Sinnarajah S. RGS2: a multifunctional regulator of G-protein signaling. *Int J Biochem Cell Biol.* 2002;34:432–438.
38. Riddle EL, Schwartzman RA, Bond M, Insel PA. Multi-tasking RGS proteins in the heart: the next therapeutic target? *Circ Res.* 2005;96:401–411. http://dx.doi.org/10.1161/01.RES.0000158287.49872.4e.
39. Timofeeva AV, Goriunova LE, Khaspekov GL, et al. Comparative transcriptome analysis of human aorta atherosclerotic lesions and peripheral blood leukocytes from essential hypertension patients. *Kardiologiia.* 2009;49:27–38.
40. Heximer SP, Knutsen RH, Sun X, et al. Hypertension and prolonged vasoconstrictor signaling in RGS2-deficient mice. *J Clin Invest.* 2003;111:445–452. http://dx.doi.org/10.1172/JCI15598.
41. Bondjers C, Kalen M, Hellstrom M, et al. Transcription profiling of platelet-derived growth factor-B-deficient mouse embryos identifies RGS5 as a novel marker for pericytes and vascular smooth muscle cells. *Am J Pathol.* 2003;162:721–729. http://dx.doi.org/10.1016/S0002-9440(10)63868-0.
42. Albig AR, Schiemann WP. Identification and characterization of regulator of G protein signaling 4 (RGS4) as a novel inhibitor of tubulogenesis: RGS4 inhibits mitogen-activated protein kinases and vascular endothelial growth factor signaling. *Mol Biol Cell.* 2005;16:609–625. http://dx.doi.org/10.1091/mbc.E04-06-0479.
43. Wang X, Adams LD, Pabon LM, et al. RGS5, RGS4, and RGS2 expression and aortic contractibility are dynamically co-regulated during aortic banding-induced hypertrophy. *J Mol Cell Cardiol.* 2008;44:539–550. http://dx.doi.org/10.1016/j.yjmcc.2007.11.019.
44. Takimoto E, Koitabashi N, Hsu S, et al. Regulator of G protein signaling 2 mediates cardiac compensation to pressure overload and antihypertrophic effects of PDE5 inhibition in mice. *J Clin Invest.* 2009;119:408–420. http://dx.doi.org/10.1172/JCI35620.
45. Zhang P, Su J, King ME, Maldonado AE, Park C, Mende U. Regulator of G protein signaling 2 is a functionally important negative regulator of angiotensin II-induced

cardiac fibroblast responses. *Am J Physiol Heart Circ Physiol.* 2011;301:H147–H156. http://dx.doi.org/10.1152/ajpheart.00026.2011.
46. Hercule HC, Tank J, Plehm R, et al. Regulator of G protein signalling 2 ameliorates angiotensin II-induced hypertension in mice. *Exp Physiol.* 2007;92:1014–1022. http://dx.doi.org/10.1113/expphysiol.2007.038240.
47. Osei-Owusu P, Sabharwal R, Kaltenbronn KM, et al. Regulator of G protein signaling 2 deficiency causes endothelial dysfunction and impaired endothelium-derived hyperpolarizing factor-mediated relaxation by dysregulating Gi/o signaling. *J Biol Chem.* 2012;287:12541–12549. http://dx.doi.org/10.1074/jbc.M111.332130.
48. Tang KM, Wang GR, Lu P, et al. Regulator of G-protein signaling-2 mediates vascular smooth muscle relaxation and blood pressure. *Nat Med.* 2003;9:1506–1512. http://dx.doi.org/10.1038/nm958.
49. Sun X, Kaltenbronn KM, Steinberg TH, Blumer KJ. RGS2 is a mediator of nitric oxide action on blood pressure and vasoconstrictor signaling. *Mol Pharmacol.* 2005;67:631–639. http://dx.doi.org/10.1124/mol.104.007724.
50. Hill MA, Meininger GA. Arteriolar vascular smooth muscle cells: mechanotransducers in a complex environment. *Int J Biochem Cell Biol.* 2012;44:1505–1510. http://dx.doi.org/10.1016/j.biocel.2012.05.021.
51. Kauffenstein G, Laher I, Matrougui K, Guerineau NC, Henrion D. Emerging role of G protein-coupled receptors in microvascular myogenic tone. *Cardiovasc Res.* 2012;95:223–232. http://dx.doi.org/10.1093/cvr/cvs152.
52. Beech DJ. Emerging functions of 10 types of TRP cationic channel in vascular smooth muscle. *Clin Exp Pharmacol Physiol.* 2005;32:597–603. http://dx.doi.org/10.1111/j.1440-1681.2005.04251.x.
53. Kim EC, Ahn DS, Yeon SI, Lim M, Lee YH. Epithelial Na+ channel proteins are mechanotransducers of myogenic constriction in rat posterior cerebral arteries. *Exp Physiol.* 2012;97:544–555. http://dx.doi.org/10.1113/expphysiol.2011.062232.
54. Chao JT, Davis MJ. The roles of integrins in mediating the effects of mechanical force and growth factors on blood vessels in hypertension. *Curr Hypertens Rep.* 2011;13:421–429. http://dx.doi.org/10.1007/s11906-011-0227-6.
55. Mederos y Schnitzler M, Storch U, Gudermann T. AT1 receptors as mechanosensors. *Curr Opin Pharmacol.* 2011;11:112–116. http://dx.doi.org/10.1016/j.coph.2010.11.003.
56. Grayson TH, Ohms SJ, Brackenbury TD, et al. Vascular microarray profiling in two models of hypertension identifies caveolin-1, Rgs2 and Rgs5 as antihypertensive targets. *BMC Genomics.* 2007;8:404. http://dx.doi.org/10.1186/1471-2164-8-404.
57. Calo LA, Pagnin E, Davis PA, et al. Increased expression of regulator of G protein signaling-2 (RGS-2) in Bartter's/Gitelman's syndrome. A role in the control of vascular tone and implication for hypertension. *J Clin Endocrinol Metab.* 2004;89:4153–4157. http://dx.doi.org/10.1210/jc.2004-0498.
58. Feletou M, Vanhoutte PM. The alternative: EDHF. *J Mol Cell Cardiol.* 1999;31:15–22. http://dx.doi.org/10.1006/jmcc.1998.0840.
59. Feletou M, Vanhoutte PM. EDHF: an update. *Clin Sci (Lond).* 2009;117:139–155. http://dx.doi.org/10.1042/CS20090096.
60. Gurley SB, Griffiths RC, Mendelsohn ME, Karas RH, Coffman TM. Renal actions of RGS2 control blood pressure. *J Am Soc Nephrol.* 2010;21:1847–1851. http://dx.doi.org/10.1681/ASN.2009121306.
61. Zuber AM, Singer D, Penninger JM, Rossier BC, Firsov D. Increased renal responsiveness to vasopressin and enhanced V2 receptor signaling in RGS2 −/− mice. *J Am Soc Nephrol.* 2007;18:1672–1678. http://dx.doi.org/10.1681/ASN.2007010032.
62. Semplicini A, Lenzini L, Sartori M, et al. Reduced expression of regulator of G-protein signaling 2 (RGS2) in hypertensive patients increases calcium mobilization and ERK1/2 phosphorylation induced by angiotensin II. *J Hypertens.* 2006;24:1115–1124. http://dx.doi.org/10.1097/01.hjh.0000226202.80689.8f.

63. Zhang J, Li N, Yan Z, et al. Association between the regulator of G-protein signaling 2 gene 1891-1892del TC polymorphism with hypertension in Xinjiang Kazakh population. *Zhonghua Yi Xue Yi Chuan Xue Za Zhi.* 2010;27:29–33. http://dx.doi.org/10.3760/cma.j.issn. 1003-9406.2010.01.006.
64. Zhao Q, Wang L, Yang W, et al. Interactions among genetic variants from contractile pathway of vascular smooth muscle cell in essential hypertension susceptibility of Chinese Han population. *Pharmacogenet Genomics.* 2008;18:459–466. http://dx.doi.org/10.1097/FPC.0b013e3282f97fb2.
65. Deschenes G, Feldmann D, Doucet A. Primary molecular changes and secondary biological problems in Bartter and Gitelman syndrome. *Arch Pediatr.* 2002;9:406–416.
66. Knoers NV, Levtchenko EN. Gitelman syndrome. *Orphanet J Rare Dis.* 2008;3:22. http://dx.doi.org/10.1186/1750-1172-3-22.
67. Katz VL. Physiologic changes during normal pregnancy. *Curr Opin Obstet Gynecol.* 1991;3:750–758.
68. Ouzounian JG, Elkayam U. Physiologic changes during normal pregnancy and delivery. *Cardiol Clin.* 2012;30:317–329. http://dx.doi.org/10.1016/j.ccl.2012.05.004.
69. Wong AY, Kulandavelu S, Whiteley KJ, Qu D, Langille BL, Adamson SL. Maternal cardiovascular changes during pregnancy and postpartum in mice. *Am J Physiol Heart Circ Physiol.* 2002;282:H918–H925. http://dx.doi.org/10.1152/ajpheart.00641.2001.
70. Clapp 3rd JF, Capeless E. Cardiovascular function before, during, and after the first and subsequent pregnancies. *Am J Cardiol.* 1997;80:1469–1473.
71. Slangen BF, Out IC, Verkeste CM, Peeters LL. Hemodynamic changes in early pregnancy in chronically instrumented, conscious rats. *Am J Physiol.* 1996;270:H1779–H1784.
72. Chu ZM, Beilin LJ. Nitric oxide-mediated changes in vascular reactivity in pregnancy in spontaneously hypertensive rats. *Br J Pharmacol.* 1993;110:1184–1188.
73. Conrad KP, Morganelli PM, Brinck-Johnsen T, Colpoys MC. The renin-angiotensin system during pregnancy in chronically instrumented, conscious rats. *Am J Obstet Gynecol.* 1989;161:1065–1072.
74. Shah DM. Role of the renin-angiotensin system in the pathogenesis of preeclampsia. *Am J Physiol Renal Physiol.* 2005;288:F614–F625. http://dx.doi.org/10.1152/ajprenal.00410.2003.
75. Friedman SA, Lubarsky SL, Ahokas RA, Nova A, Sibai BM. Preeclampsia and related disorders. Clinical aspects and relevance of endothelin and nitric oxide. *Clin Perinatol.* 1995;22:343–355.
76. Kvehaugen AS, Melien O, Holmen OL, et al. Single nucleotide polymorphisms in G protein signaling pathway genes in preeclampsia. *Hypertension.* 2013;61:655–661. http://dx.doi.org/10.1161/HYPERTENSIONAHA.111.00331.
77. Kvehaugen AS, Melien O, Holmen OL, Laivuori H, Dechend R, Staff AC. Hypertension after preeclampsia and relation to the C1114G polymorphism (rs4606) in RGS2: data from the Norwegian HUNT2 study. *BMC Med Genet.* 2014;15:28. http://dx.doi.org/10.1186/1471-2350-15-28.
78. Gosens R, Zaagsma J, Meurs H, Halayko AJ. Muscarinic receptor signaling in the pathophysiology of asthma and COPD. *Respir Res.* 2006;7:73. http://dx.doi.org/10.1186/1465-9921-7-73.
79. Jiang H, Xie Y, Abel PW, et al. RGS2 repression exacerbates airway hyperresponsiveness and remodeling in asthma. *Am J Respir Cell Mol Biol.* 2014. http://dx.doi.org/10.1165/rcmb.2014-0319OC.
80. Holden NS, Bell MJ, Rider CF, et al. beta2-adrenoceptor agonist-induced RGS2 expression is a genomic mechanism of bronchoprotection that is enhanced by glucocorticoids. *Proc Natl Acad Sci USA.* 2011;108:19713–19718. http://dx.doi.org/10.1073/pnas.1110226108.

CHAPTER SIX

Keeping the Balance Right: Regulator of G Protein Signaling 5 in Vascular Physiology and Pathology

Ruth Ganss[1]
Harry Perkins Institute of Medical Research, Centre for Medical Research, The University of Western Australia, Perth, Western Australia, Australia
[1]Corresponding author: e-mail address: ganss@perkins.uwa.edu.au

Contents

Abstract

The cardiovascular system including the heart and intricate network of arteries, veins, and capillaries is essential for normal organ homeostasis and tightly controlled by G protein-coupled receptor (GPCR) signaling cascades. Imbalances of these signaling systems can manifest in cardiovascular disease. There has been a recent surge in studies on modulators of GPCR activity, so-called regulator of G protein signaling (RGS)

Progress in Molecular Biology and Translational Science, Volume 133
ISSN 1877-1173
http://dx.doi.org/10.1016/bs.pmbts.2015.02.003

molecules, due to their potential as pharmacological targets. Among RGS proteins, RGS5 is prominently expressed in arterial vascular smooth muscle cells (vSMC) suggesting an important role in vascular function. Although apparently dispensable for embryonic development, RGS5 has now emerged as a crucial regulator of adaptive cardiovascular processes, including remodeling of the vascular wall under stress. RGS5 has been shown to regulate signaling pathways which shape vSMC differentiation, migration, contraction, as well as tissue inflammation and fibrosis. Indeed, studies in RGS5 mutant mice have confirmed a crucial and nonredundant role as regulator of cardiac function, blood pressure homeostasis, and adult neovascularization such as angiogenesis and arteriogenesis. In response to environmental cues, RGS5 is dynamically controlled at both mRNA and protein levels. This enables direct, precise, and rapid modulation of Gαq- and Gαi-coupled GPCR signaling which also integrates receptor tyrosine kinases (RTK) and Gαs/Gα12/13-mediated GPCR signal transduction. Although RGS5's endogenous role in a spatiotemporal context is still largely unknown, its prominence in vascular tissue makes it an important molecule to study and a prime candidate for therapeutic intervention.

1. REGULATOR OF G PROTEIN SIGNALING MOLECULES IN THE CARDIOVASCULAR SYSTEM

Signaling through GPCRs and their heterotrimeric G proteins is an integral part of the developing and adult cardiovascular system. Upon ligand binding, the Gα protein exchanges GDP for GTP and dissociates from the Gβγ unit which enables both subunits to signal autonomously. GPCR signaling is fine tuned by RGS molecules which are best known for their GTPase-activating function (GAPs). RGS proteins interact specifically with Gα subunits (Gαi/o, Gαq/11, Gαs) to accelerate their intrinsic GTPase activity resulting in timely signal termination and reassociation of Gα and Gβγ subunits.[1] Thus, by virtue of their GAP activity, RGS proteins are negative regulators of GPCR signaling. In addition, RGS molecules can bind to a range of other molecules including receptors and scaffold proteins which confer regulatory functions beyond their GAP activity.[2,3] A series of RGS family members are expressed in the cardiovascular system.[4] Early studies on isolated cardiomyocytes and vSMC provided primary evidence for a functional role of RGS proteins in cardiovascular physiology for instance by regulating the response to vasoactive hormones such as angiotensin II (AngII), endothelin-1 (ET-1), and sphingosine-1-phosphate (S1P).[5,6] More recently, mutant mice harboring a point mutations in their Gαi/o or Gαq/11 subunits that disrupts Gα-RGS binding without affecting receptor coupling demonstrated a crucial role for endogenous RGS proteins for instance in diabetic cardiomyopathy, ischemic heart failure, and vascular injury.[7–9] Given the increasing number of different RGS domain-containing proteins in mammals (>35) insights into tissue- and receptor-specific roles of individual regulators *in vivo* are urgently needed. Most RGS proteins which are strongly expressed in the cardiovascular system belong to the B/R4 subfamily. B/R4 RGS proteins (RGS 1–5, 8, 13, 16, 18, and 21) are among the smallest RGS proteins and thus offer only few potential contact sites outside the RGS domain.[10] A notable exception is RGS6 which is highly expressed in the heart but belongs to the R7 family characterized by two additional domains, the Disheveled-EGL-10-Pleckstrin (DEP) homology domain and a Gγ subunit-like domain which confer membrane association and protein stability, respectively.[11] In the last decade, considerable progress in the understanding of the physiological roles, in particular of RGS2, 4, 5, and 6, in the heart and/or vasculature has been made by using knockout and transgenic animals, and in some cases by identifying naturally occurring mutations in humans.[12–14] This review focuses on the B/R4 family member RGS5 with an emphasis on its regulatory function in the cardiovascular system and recent findings in mutant mouse models.

2. RGS5: AN EMERGING PLAYER IN THE CARDIOVASCULAR SYSTEM

Murine RGS5 was first cloned in the late 1990 and shown to inhibit GPCR-mediated pheromone signaling in yeast.[15] RGS5 exhibits strongest sequence similarities with B/R4 family members RGS4 and RGS16. Human and mouse RGS5 mRNA are 90% identical suggesting evolutionary conservation of a crucial function between species.[16] Biochemical characterization of RGS5 demonstrates preferential binding to Gαi/o and Gαq subunits but not to Gαs or Gα13.[17,18] High RGS5 levels suppress AngII and ET-1-induced intracellular calcium (Ca^{2+}) signaling concordant with attenuation of Gαq-mediated activation of phospholipase C and reduced inositoltrisphosphate (IP3) production; this action is dependent on the RGS domain since the N-terminal 33 amino acids which flank the RGS domain are dispensable for its GAP function[17] (Fig. 1A). Consistently, specific knockdown of endogenous RGS5 in rat vSMC enhances mitogen-activated protein (MAP) kinase signaling downstream of the AngII receptor type 1 (ATR1) most likely via association with Gαq.[6] Earlier studies by Schwartz and colleagues identified RGS5 as the most highly expressed gene

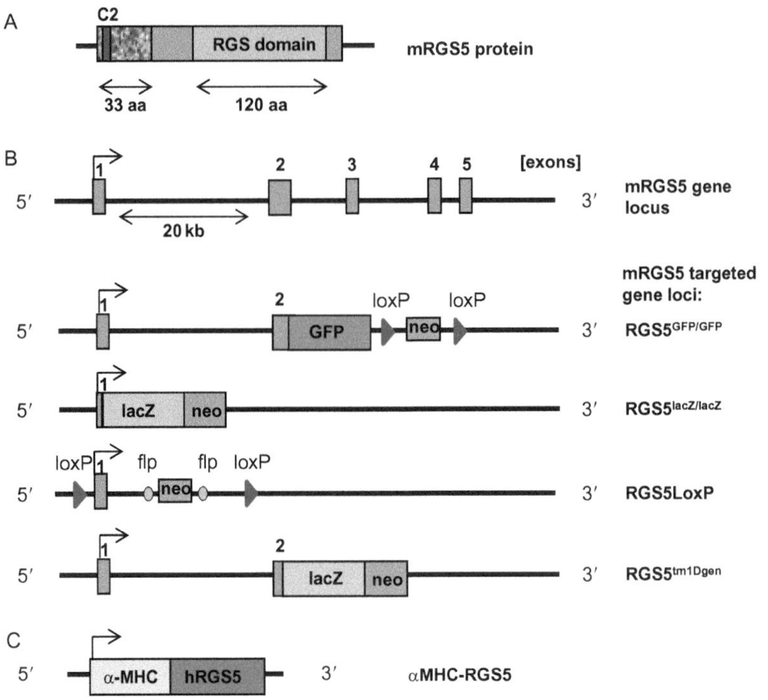

Figure 1 Murine RGS5 protein/gene structures and RGS5-targeted gene loci. (A) Murine RGS5 protein consists of a 120 aa conserved RGS domain which confers GTPase activity and a 33 aa N-terminus which contains a cysteine residue at position 2 (C2) and a potential amphipathic helix domain. (B) The murine RGS5 gene comprises 5 exons. Gene deletion analysis demonstrates that intron 1 (approximately 20 kb) contains important regulatory elements[19] and a functional PPARγ and β/δ binding site.[20] RGS5 knockout mouse strains were generated using the following strategies: Nisancioglu *et al.* inserted a green fluorescent protein (GFP) and a neomycin resistance gene flanked with LoxP sites in frame with exon 2. The selection cassette (neo) was removed using Cre/LoxP technology (RGS5$^{GFP/GFP}$).[21] Cho *et al.* generated knockout mice by cloning a lacZ/neomycin gene cassette (LacZ neo) in frame with exon 1. This strategy produced RGS5-deficient mice, but the lacZ gene was inactive and the neo cassette was not removed (RGS5$^{LacZ/LacZ}$).[19] Hamzah *et al.* flanked exon 1 with LoxP sites. The neo cassette was removed by flp recombination (RGS5LoxP). RGS5-deficient mice were generated by crossing RGS5-LoxP mice with Cre-deleter mice.[22] Another LacZ/neo expressing RGS5 knockout strain was generated by Deltagen Inc. In this line, lacZ/neo was inserted into exon 2 and is expressed from the endogenous RGS5 promoter (RGS5^{tm1Dgen}). (C) A cardiac myocyte-specific, RGS5 overexpressing mouse line was generated by cloning the human cardiac myosin heavy chain (αMHC) promoter in front of the human RGS5 cDNA (αMHC-RGS5).[23] (See the color plate.)

in a comparison between aortic and vena cava vSMC, implying a crucial role in aortic function.[24,25] Subsequently, Betsholtz and colleagues identified RGS5 as specific marker for mural cells, a cell population consisting of pericytes and vSMC which line and support endothelial cells. Specifically, platelet-derived growth factor B (PDGFB) and PDGFB receptor-deficient mice lack pericytes and RGS5 expression in the microvasculature.[26] Moreover, vascular RGS5 expression is developmentally regulated in the mouse embryo and peaks at around E14.5 followed by a steady decline.[27] In adult mice, RGS5 arterial expression persists, but shows specific differences between vascular beds, reflecting the developmental heterogeneity of vSMC. This has been shown in a RGS5-LacZ reporter mouse ($RGS5^{tm1Dgen}$; Fig. 1B) where weak expression in carotid arteries contrasts strong RGS5 promoter activity in the descending aorta and renal arteries. Interestingly, regional differences in RGS5 expression may be governed by promoter methylation and epigenetic mechanisms.[28] Thus, RGS5 expression levels correlate with vessel development and maturation, but its specific role in these processes still remains unclear. Surprisingly, vascular beds in RGS5-deficient mice develop normally with no defects in pericyte coverage.[28] Nevertheless, RGS5 has frequently been detected in gene profiling experiments as dynamically regulated in vascular pathologies indicative of a role in adaptive processes in the adult.[25,29–32] Indeed, while developmental neovascularization may not depend on RGS5, or redundancy by other family members can compensate for early loss of RGS5, there is strong evidence for a key function for RGS5 in cardiovascular physiology and pathology.

3. RGS5 AND CARDIOVASCULAR PATHOLOGIES

3.1 A Cardioprotective Role of RGS5

Similar to vSMC, cardiac myocyte growth and contraction is crucially regulated by GPCR signaling and indeed several RGS proteins, in particular of the B/R4 family, are expressed in ventricular myocytes and fibroblasts.[14,33] Moreover, a role for RGS2 and RGS4 in cardiac pathology has been well established in mutant mice.[34,35] RGS5 expression is induced in atria under chronic β adrenergic stimulation suggesting a potential role in heart failure.[36] Indeed, loss or gain of RGS5 expression in the heart of mutant mice has dramatic functional consequences. For instance, cardiac-specific overexpression of RGS5 using the α-myosin heavy chain (MHC) promoter (Fig. 1C) protects mice from hypertrophy following prolonged aortic banding. In contrast, aortic banding in RGS5-deficient mice leads to more substantial cardiac hypertrophy than in wild-type controls suggestive of a major role of RGS5 in cardiac adaptation following pressure overload.[23] Loss of RGS5 increases MEK/ERK signaling in hypertrophic hearts which can be reversed with specific pathway inhibition. Moreover, aortic banding in $RGS5^{-/-}$ mice induces perivascular and interstitial fibrosis due to enhanced TGF-β/Smad signaling. Similarly, collagen synthesis is stimulated in isolated cardiac fibroblasts in the absence of RGS5. *In vivo* inhibitor studies indicate that enhanced TGFβ signaling is linked to MEK/ERK activation consistent with synergistic promotion of collagen synthesis by both pathways.[37] Interestingly, loss of RGS5 in the two-kidney one-clip model of chronic hypertension also aggravates vascular damage and causes hypertrophy of renal resistance arteries and perivascular fibrosis in small renal arteries. While MEK/ERK signaling inhibition resolves hypertrophic arterial growth, it is Rho kinase inhibition which reduces perivascular fibrosis.[38] These studies demonstrate that RGS5 regulates both, hypertrophic vessel remodeling and fibrosis, possibly involving distinct pathways.

3.2 RGS5 Regulates Vascular Tone and Blood Pressure

In healthy adults, RGS5 is strongly expressed in peripheral arterial smooth muscle cells.[24,28] Moreover, RGS5 inhibits signaling of the potent vasoconstrictor AngII in isolated vSMC and aortic rings in a Gαi- and Gαq-dependent manner.[6,39] RGS2 and RGS5 are downregulated in saphenous arteries from hypertensive rats suggesting a correlation between reduced RGS expression and hypertension.[32] In humans, gene polymorphisms in both, RGS2 and RGS5, have been associated with hypertension.[40–44] Indeed, RGS2 deficiency in mice has been shown to cause hypertension even though the magnitude varies between experiments.[45–47] Based on biochemical data, RGS5 knockout mice are expected to be hypertensive.[6] However, to this end, RGS5-deficient mice have been reported to be hypotensive,[19,21] normotensive,[48] or hypertensive[38] (Fig. 1B and Table 1). Differences in knockout strategies, genetic background, and/or blood pressure assessment methods may in part explain the discrepancies. The most comprehensive blood pressure study to date demonstrates that loss of RGS5 causes hypertension.[38] Consistent with previous findings in RGS5 overexpressing cells,[6] AngII stimulation of primary RGS5-deficient vSMC enhances intracellular Ca^{2+} levels and myosin light chain (MLC) activation compared to wild-type cells. Furthermore, hypertension in RGS5-deficient

Table 1 Cardiovascular Phenotypes in Different RGS5 Knockout Mice

Mouse Model	Cardiovascular Phenotype in RGS5-Deficient Mice	Method	Genetic Background	References
$RGS5^{GFP/GFP}$	• Hypotension • Retinopathy (no vascular phenotype) • Tumor angiogenesis (no vascular phenotype) • Arteriogenesis (reduced growth of collateral arteries). • Reduced neointima formation. • Normotension	• Tail cuff (MAP) • Oxygen-induced/IHC • s.c. fibrosarcoma/IHC • Femoral artery ligation/ Doppler analysis. • Left carotid artery ligation, IHC. • Radiotelemetry	C57BL/6	21 48
$RGS5^{LacZ/LacZ}$	• Hypotension. • No cardiac phenotype • Dilated aorta	• Tail cuff (MAP) • Electrocardiography • Histology	129Sv × C57BL/6	19
RGS5LoxP × Cre-deleter ($RGS5^{-/-}$)	• Tumor angiogenesis (reduced leakiness, enhanced vessel maturation) • Hypertension	• MRI/IHC • Radiotelemetry (MAP), Millar probe.	C3HeBFe or C57BL/6	38 22
$RGS5^{tm1Dgen}$	• Cardiac hypertrophy • Normotension. • Prolonged cardiac repolarization, prolonged atrial and ventricular tachyarrhythmia.	• Aortic banding, echocardiography/IHC • Millar probe (SBP). • Radiotelemetry ECG • Electrophysiology on isolated hearts.	C57BL/6	23 49 50
αMHC-hRGS5	• Protection from cardiac hypertrophy	• Aortic banding, echocardiography/IHC	C57BL/6	23

mice is associated with arterial hypercontractility, increased vascular resistance, and vessel stiffening which leads to a more severe hypertensive phenotype in aged RGS5-deficient mice. In the context of blood pressure regulation, arterial RGS5 controls multiple pathways downstream of AngII which includes MEK/ERK and Rho kinase signaling; these pathways act cooperatively to increase blood pressure in the absence of RGS5 and specific inhibitors of both pathways reduce blood pressure to normal wild-type levels.[38] Thus, RGS5 has a critical and nonredundant role in vascular homeostasis and blood pressure regulation. So far, mechanistic studies have focused on peripheral vascular resistance. Given the prominent expression of RGS5 in nervous system, heart, and kidney, it is likely that elevated blood pressure is also influenced by other mechanisms such as parasympathetic and sympathetic control as reported for RGS2.[51]

3.3 A Critical Role of Vascular RGS5 in Airway Hyperresponsiveness

Airway or bronchial smooth muscle cells (aSMC) are close relatives of vSMC and share many structural and biochemical features, and importantly, play a prominent role in the pathogenesis of asthma.[52] Functional RGS5 studies in aSMC have significantly advanced the cardiovascular field and are therefore discussed in the context of this review. Inhalation treatment with β_2-agonists provides temporary relief for asthmatic patients by dilating airways via the $G\alpha_s$-cAMP axis. However, prolonged treatment, in particular without simultaneous intake of steroids, ultimately leads to airway hypersensitivity and bronchoconstriction. While several potential mechanisms for this observation have been described which include downregulation of Gαs[53] or upregulation of contractile receptors,[54] Druey and colleagues discovered that RGS5 (and RGS4) is downregulated in chronically stimulated aSMC, independent of GPCR levels.[55] This in turn leads to stronger intracellular Ca^{2+} flux, MLC phosphorylation, and contraction most likely due to enhanced Gαq mediated pro-contractile signaling;[55] this is reminiscent of *in vitro* RGS knockdown studies in vSMC and *in vivo* regulation of vascular tone in RGS5-deficient mice.[38] In addition, precision cut lung slices from RGS5 knockout mice contract stronger upon application of muscarinic receptor ligands such as carbachol.[55] Interestingly, RGS5 gene polymorphism correlates with clinical response to β-agonists in asthmatic children providing additional support for a regulatory role of RGS5 in airway hypersensitivity.[56] However, at this stage, it is unclear whether RGS5 regulates normal lung function or the healing process after injury. In contrast to chronic bronchodilation, RGS5 is overexpressed in patients suffering from a severe form of asthma associated with aSMC hypertrophy. In these patients, elevated RGS5 expression impairs intracellular Ca^{2+} flux and reduces contraction.[57] Similarly, in a model of suprarenal abdominal aortic banding RGS5 expression is significantly upregulated in thoracic aorta 28 days after constriction concomitant with arterial hypertrophy and attenuated contraction.[39] Thus, imbalance in RGS5 expression, be it up- or downregulation, can cause pathology in a highly context-dependent manner.

3.4 RGS5 Transcript Levels Correlate with Severity of Atherosclerosis

RGS5 is predominantly expressed in medial SMC of arteries; however, RGS5 mRNA expression is reduced in the arterial neointima, e.g., in bypass graft neointima and atherosclerotic arteries in monkeys, compared to the normal arterial wall.[25,30] Moreover, the fibrous cap of advanced atherosclerotic lesions in humans has considerably less RGS5 mRNA than non-atherosclerotic intima.[58] While this suggests a potential involvement of RGS5 in vSMC phenotypic switching during vascular injury, support for a functional role of RGS5 in plaque development is still lacking. Some indirect evidence comes from an inflammatory model of enhanced atherosclerosis where low-density lipoprotein receptor (LDLR) knockout mice were fed a high-fat diet and simultaneously infused with AngII.[59] In this model, RGS4 and RGS5 transcripts in the aorta are downregulated with disease progression implying that AngII amplifies its own signaling by down-regulating RGS molecules, as has been observed in other models.[38] Peroxisome proliferator-activated receptor (PPAR)δ agonists have a protective effect by reducing vascular inflammation and macrophage recruitment into lesions which also coincides with upregulation of RGS4 and 5 transcripts. Similarly, in an atherosclerotic disease model of cholesterol-fed apolipoprotein (Apo)E-deficient mice, PPARδ agonist treatment reduces plaque size concomitant with a reduction in inflammatory cytokines/chemokines, and upregulation of RGS4 and RGS5 mRNA.[60] Furthermore, when ApoE-deficient mice on a high-fat diet are treated with lipid lowering and anti-inflammatory drugs of the statin family, atherosclerotic plaque formation in the aorta is reduced.[61] Interestingly, among many other genes, RGS5 mRNA is upregulated in this study indicating that statins and PPARδ agonists increase RGS5 expression in aorta which in turn may regulate hypertrophic and/or inflammatory signaling upon injury. Incidentally,

RGS5 knockout mice on a high-fat diet are prone to become more obese than wild-type littermates and also show increased JNK and NF-κB stimulation and inflammation.[62] Taken together, these studies strongly support a role of RGS5 in inflammation; at present, it is unclear, however, whether RGS5 plays a causative role in plaque progression or is merely a marker for pathological vascular remodeling.[a] In this context, it is of interest that the activity of the RGS5 gene promoter is controlled by methylation.[28] Recently, it has been shown that disturbed blood flow in arteries epigenetically modulates endothelial gene transcription by upregulation of DNA methyltransferase which ultimately enhances atherosclerosis.[63] It is conceivable that RGS5 promoter methylation changes under vascular flow, not dissimilar to RGS5 transcriptional regulation under hydrostatic pressure.[64]

3.5 Tumor Angiogenesis: The Less RGS5 the Better

In adults, neovascularization is part of normal tissue remodeling for instance in female reproductive organs and formation of granulation tissue during wound healing. Sprouting of existing blood vessels in response to angiogenic factors follows a tightly regulated program which involves pericyte detachment from vessel walls which in turn facilitates endothelial cell migration and formation of new tubes. Pericyte reattraction into vascular beds in response to growth factors such as PDGFB then helps to mature and stabilize the nascent endothelial sprouts.[65] Consistent with vascular remodeling in the embryo, RGS5 expression is temporarily upregulated in blood vessels during ovulation and tissue repair.[31] Uncontrolled blood vessel sprouting, for instance during tumor angiogenesis, creates profoundly abnormal vessels with a typical chaotic, dilated, and hyperpermeable phenotype not least because of reduced pericyte-endothelial cell alignment and/or pericyte maturity.[66,67] During tumor angiogenesis, the transition from prevascular hyperplasia to highly vascularized tumors is termed the "angiogenic switch."[68,69] Interestingly, RGS5 is upregulated concomitant with the earliest vascular abnormalities at the time of the angiogenic switch and subsequently highly expressed in angiogenic vessels of solid tumors in the RIP-Tag (RIP, rat insulin gene promoter; Tag, SV40 Large T antigen) mouse model of pancreatic neuroendocrine cancer. Similarly, angiogenic neovessels in brain tumors are strongly positive for RGS5; this is in contrast to brain tumors in HIF1α-knockout mice which—due to defects in the neovascularization process—grow by co-opting existing vessels. Moreover,

[a] Cheng and colleagues have recently shown that RGS5 deficiency enhances plaques instabillity in $RGS5^{-/-}$ $ApoE^{-/-}$ mice.[138]

RGS5 expression is drastically reduced in tumor vessels which normalize under therapy.[31] These studies establish RGS5 as a pericyte-specific marker for pathological neovascularization which is potentially upregulated during the phase of pericyte recruitment to newly formed vascular sprouts.[70] To answer the question whether RGS5 is merely a marker for vessel remodeling or serves a specific function during pathological adaptation, RIP-Tag tumor-bearing mice were crossed onto a RGS5-deficient background. Interestingly, cancers in these mice still develop but with a less chaotic and leaky vasculature, attachment of more mature pericytes, and improved vessel function[22] (Fig. 2). As a consequence, RGS5$^{-/-}$ tumors are less hypoxic and, importantly, highly responsive to anticancer immune therapy.[12,22,71] This clearly demonstrates that RGS5 actively regulates vascular barriers in tumors. Moreover, in a model of transient cerebral ischemia RGS5 deficiency reduces vessel permeability and brain edema[22] consistent with a role of pericytes in regulation of the blood–brain barrier and an essential role of RGS5 therein.[72] How RGS5 expression regulates pericyte maturity/vessel coverage, vascular barrier integrity, and immune cell infiltration into tumors is at present unknown. Recently, pericytes have been implicated in controlling innate immune cell migration and function in the periphery.[73,74] Therefore, it is conceivable that

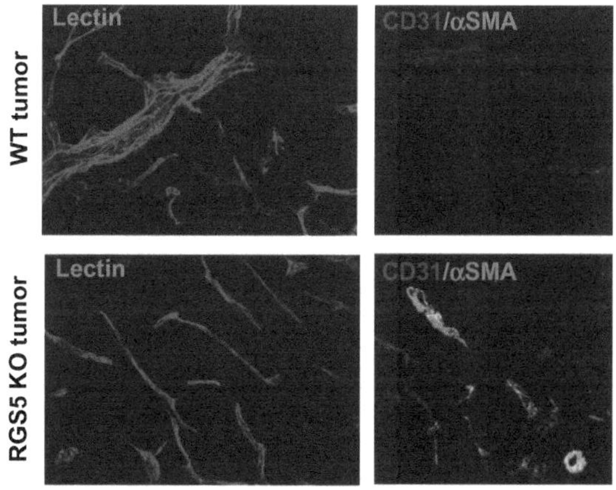

Figure 2 RGS5-deficient tumor vessels are normalized. Confocal images showing tumor vessels in a fibrosarcoma grown in wild-type mice. Lectin-FITC perfused vessels with typical heterogeneous vessel caliber are depicted (upper left panel). In contrast, vessels from tumors grown in RGS5 knockout mice are homogeneous in caliber (lower left panel, magnification 20 ×). Blood vessels (CD31 staining in red) in wild-type tumors are immature with few αSMA-positive pericytes (αSMA staining in green, upper right panel). Normalized blood vessels in RGS5 knockout mice show increased coverage by αSMA-positive pericytes which stabilize vessels (lower left panel, magnification 60 ×).[22] (See the color plate.)

RGS5 actively controls effector T cell access into malignant tissue. This is supported by data demonstrating that intratumoral $RGS5^+$ pericytes can be immunosuppressive and may thus play a role in antitumor immunity.[75] In human cancer, RGS5 has been localized to the angiogenic vasculature for instance in hepatocellular carcinoma, renal cell carcinoma, ovarian, and gastric cancer where it seems to be expressed in endothelial cells rather than pericytes.[76–80] RGS5 has also been identified as a potential tumor antigen which suggests that it is expressed in tumor cells of multiple different cancer types.[81–84] Moreover, RGS5 is part of a three gene signature which predicts survival in patients with renal cell carcinoma.[85] Most recently, polymorphisms in genes involved in pericyte maturation, including RGS5, have been shown to be of potential predictive value for bevacizumab (antivascular endothelial growth factor-A (VEGF-A))/chemotherapy combination treatment in colorectal cancer.[86]

Pericytes are a heterogeneous population and in addition to their role in supporting the microvasculature they appear to have stem cell-like properties.[87,88] Interestingly, RGS5 is expressed in human dermal pericytes which promote skin regeneration by modulation of extracellular matrix in a process which is surprisingly independent of neovascularization.[89] RGS5 is also overexpressed, both perivascularly and scattered throughout the dermis, in scleroderma, an autoimmune connective tissue disease with distinct vascular and fibrotic pathology.[90] Moreover, hepatic stellate cells, a pericyte-like cell type, upregulate RGS5 in response to injury and loss of RGS5 results in exacerbated fibrosis.[91] These findings indicate that RGS5 may be a marker for pericytes which regulate stromal remodeling and fibrosis in a broader context beyond vascular beds.[92]

3.6 RGS5 Regulates vSMC Phenotype During Arteriogenesis

As an important regulator of vessel wall remodeling and adult neovascularization, RGS5 is also important in the process of arteriogenesis, the collateral arterial growth which bypasses obstruction of large conduit arteries.[48] Collateral vessel growth is stimulated by shear stress as a result of increased blood flow in blocked arteries and can occur in heart, brain, kidney, and the peripheral circulation. Arteriogenesis, like angiogenesis, is tightly controlled by key signaling pathways involving nitrite oxide (NO), Rho kinase, and shear stress sensitive Ca^{2+} channels.[93] Shear stress-activated collateral endothelial cells release NO which in turn relaxes vSMC and dilates vessels. This is followed by leukocyte recruitment into the vessel wall, inflammatory vessel remodeling, and extensive vSMC proliferation.[94] Upon ligation of femoral arteries and experimental stimulation of collateral arterial growth in the hind limb, RGS5 is upregulated in vSMC concomitant with loss of myocardin expression, a marker for vSMC differentiation. Conversely, in RGS5-deficient animals, development of a collateral flow bypass is impaired, demonstrating an active role for RGS5 in this process.[48] In the absence of RGS5, differentiated and quiescent vSMC persist, fail to activate RhoA, and attract macrophages. RGS5 is upregulated *in vitro* by prolonged NO exposure of isolated vSMC or arterial fragments, as well as cyclic stretch. Elevated RGS5 levels in turn lead to reduced Ca^{2+} release and contraction in response to vasoconstrictors indicative of reduced Gαq-mediated signaling as observed in other experimental models.[6,38,39,57] Surprisingly, RGS5 induction in vSMC enhances Rho kinase-dependent stress fiber formation, which is inhibited by overexpression of the LscRho GTP exchange factor (GEF), a specific antagonist of Gα12/13 signal transduction.[95] Thus, RGS5 overexpression during collateral vessel growth appears to downregulate Gαq but stimulate Gα12/13 signaling. While RGS5 is a well-established modulator of Gαq and Gαi-mediated signaling,[6,17] this is the first report which links RGS5 to Gα12/13.[48] Although this model of experimental arteriogenesis provides new insights into RGS5 function, not all findings may be generally applicable. For instance, increased RhoA signaling in vSMC has been shown in other models to promote vSMC differentiation rather than dedifferentiation.[96] Moreover, RGS5 protein is a potential target of NO-mediated degradation,[97,98] which may lead to reduced rather than elevated RGS5 levels after NO exposure in other systems. Also, RGS5 mRNA is absent in coronary arteries from monkeys,[25] and therefore, an endogenous role for RGS5 in supporting arteriogenesis may be restricted to cerebral or peripheral ischemic arterial disease.

4. DYNAMIC REGULATION OF RGS5 LEVELS

4.1 RGS5 Transcriptional Regulation

There is ample evidence for transcriptional regulation of the RGS5 gene as part of dynamic cardiovascular adaptations following environmental stimuli. For instance, RGS5 mRNA, but not RGS2 or RGS4, is increased in HUVEC cells under hypoxia.[99] This finding is consistent with a role of RGS5 in tumor angiogenesis[31] although active HIF-1 binding sites in the RGS5 gene have not yet been identified.[99] Furthermore, *in vitro* studies demonstrate that RGS2, RGS4, and RGS5 transcripts are downregulated in vSMC after stimulation with PDGFB which consequently enhances migration toward PDGFB, GPCR signaling, and hypertrophy.[100] Even though RGS5 expression levels increase with chronic AngII infusion or repeated isoproterenol injections *in vivo*,[36,38,59] *in vitro*, there is no evidence for direct transcriptional regulation of RGS5 by GPCR agonists such as AngII, ET-1, S1P or the β adrenergic agonist isoproterenol in vSMC, or cardiomyocytes.[100,101] As discussed above, the RGS5 gene is epigenetically controlled and subject to promoter hypermethylation which correlates with vascular bed and maturation-specific expression.[28] While these studies provide insights into developmental and environmental cues which regulate RGS5 mRNA levels, until recently little has been known about direct transcriptional regulators. Interestingly, a correlation between PPARδ or PPARγ agonist treatment and RGS5 expression in rodents has been described in different experimental models. For instance, PPARδ agonists decrease vascular inflammation in an AngII-accelerated model of atherosclerosis which also prevents AngII-induced downregulation of arterial RGS4 and RGS5 mRNA.[59] Moreover, PPARδ agonists lower blood pressure in hypertensive rats concomitant with an increase in vascular RGS5 expression.[102,103] PPARs are nuclear receptor proteins and transcriptional regulators with a broad range of activities; drugs which modulate PPARs are currently in clinical use to control lipid metabolism and metabolic syndrome.[104] However, there is now increasing evidence for a prominent vascular role of PPARs.[105] Indeed, recently Ketsawatsomkron *et al.* discovered an active PPAR response element within intron 1 of the mouse RGS5 gene, which strongly supports a direct transcriptional control of RGS5 by ligand activated PPARs (Fig. 3A). Intriguingly, expression of dominant negative PPARγ specifically in blood vessels in mutant mice downregulates RGS5 expression and increases the AngII-induced pressor response in resistance arteries[20] consistent with a hypertensive phenotype in RGS5-deficient mice.[38] Thus, PPARγ and PPARδ agonists have emerged as interesting new tools for pharmacological manipulation of vascular RGS5.

4.2 RGS5 Protein Modification, Stability, and Regulation by the N-end Rule Pathway

RGS5 as a member of the B/R4 RGS family is relatively small and the approximately 120 aa core RGS domain dominates its structure and function (Fig. 1A). Nevertheless, RGS4, RGS5, and RGS16 share a 33 aa N-terminal domain with significant sequence homology, indicative of a

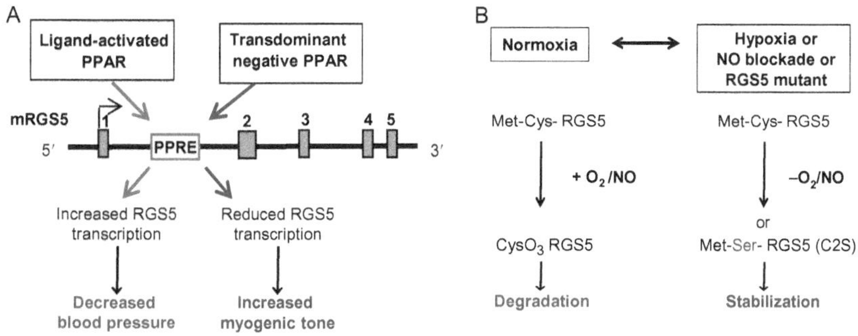

Figure 3 Schematic representation of transcriptional and proteasomal regulation of RGS5. (A) The RGS5 gene contains an active PPAR response element (PPRE) within intron 1.[106] Ligand-activated PPARβ/δ induces RGS5 mRNA in aorta of spontaneously hypertensive rats concomitant with a reduction in blood pressure.[102] Overexpression of transdominant negative PPARγ specifically in blood vessels decreases RGS5 mRNA and increases myogenic tone in mice.[106] (B) Posttranslational control of RGS5 levels. The N-terminal cysteine residue (Cys, C2) in the RGS5 protein is essential for the ubiquitin-proteasome-dependent N-end rule degradation pathway. Specifically, NO-mediated oxidation of N-terminal Cys ($CysO_3$) is required for its arginylation by arginyl transferase (ATE-1). Arginylation then leads to degradation by E3 ubiquitin ligases. Mutation of Cys to Ser at position 2 in RGS5 stabilizes the protein. Met, methionine; Cys, cysteine; Ser, serine; O_2, oxygen; NO, nitric oxide. *Modified from Ref. 98.* (See the color plate.)

conserved function. Formation of amphipathic α helical structures in this region confers membrane association as demonstrated for RGS4 and RGS16.[107,108] Interestingly, RGS5, along with RGS4 and 16, has been identified as an *in vivo* substrate of the N-end rule protein degradation pathway with requires the presence of an N-terminal cysteine (at position 2, C2).[97,98,109] The conjugation of arginine, by arginyl transferase (ATE1), to N-terminal, oxidized C2 is required for recognition by ubiquitin ligases and subsequent proteasomal degradation (Fig. 3B). In short, the N-end rule pathway acts as an environmental sensor which regulates GPCR signaling through the O_2/NO-controlled degradation of RGS4, 5, and 16. Intriguingly, ATE1 knockout embryos which are deficient in N-end degradation, die of cardiovascular defects, including ventricular hypoplasia and impaired late angiogenesis. This suggests that the ATE1-dependent proteolysis is a crucial mechanism for myocardial growth and blood vessel integrity. Recently, Lee *et al.* demonstrated that ATE-1-deficient embryonic hearts and cardiomyocytes but not cardiac fibroblasts show impaired PKC-MEK/ERK signaling which can be rescued by Gαq overexpression.[110] In this study, the authors postulate that RGS proteins may be continuously degraded under normal physiological conditions to ensure maximal responsiveness to environmental cues. Decline of O_2 under ischemic or hypoxic conditions would then lead to RGS accumulation, reduction in GPCR signaling, and decoupling from extracellular proliferation signals. So far, evidence for a functional role of the N-end rule *in vivo* has been demonstrated for RGS4 in the context of angiogenesis and hypertrophy in adult hearts.[111] Interestingly, cysteine 2 (C2) in RGS4 is also a palmitoylation site and covalent attachment of palmitoyl groups *in vivo* interferes with proteasomal degradation, thus prolonging protein half life in a remarkable convergence of pathways at the C2 residue.[112] Currently, there is little *in vivo* evidence for a physiological role of N-end-mediated RGS5 degradation or palmitoylation. It is interesting to speculate, however, that enhanced brain edema in adult ATE1-deficient mice[113] might be linked to pathological RGS5 stabilization since the opposite, that is, improved brain edema and barrier function, is observed when RGS5 is lost.[22]

NO is also a potent vasodilator and thus plays a key role in regulation of vascular tone. Intriguingly, both RGS2 and RGS5, regulate blood pressure homeostasis in a nonredundant way. Only RGS5 is a direct target for the NO-mediated N-end rule degradation pathway and a functional role of this pathway in blood pressure regulation remains to be shown. RGS2 activity is indirectly regulated by NO via activation of cGMP and cGMP-dependent protein kinase G (PKG)1α which phosphorylates and activates RGS2 (at serine 64). This in turn leads to inhibition of GPCR-mediated vasoconstrictor signaling and enhanced vessel dilation.[45] Interestingly, while RGS5 has a potential PKC target site (at serine 16),[114] PKG1α-phosphorylation sites are absent which may in part explain the nonredundant physiological roles of RGS2 and RGS5 in blood pressure control.

5. RGS5 AND vSMC PHENOTYPIC PLASTICITY: JACK OF ALL TRADES, MASTER OF NONE?

5.1 RGS5 Links GPCR and Receptor Tyrosine Kinase Signaling

In vitro and *in vivo* studies demonstrate a key role for RGS5 in vSMC contraction and pathological vascular adaptation. Vascular SMC and pericytes, much as RGS5 expression itself, are dynamically regulated and undergo phenotypic switching in response to physiological and pathological stimuli.[115] Indeed, it is conceivable that RGS5 is causally involved in vSMC phenotypic plasticity. Acute vascular injury induces vSMC dedifferentiation into a synthetic phenotype characterized by enhanced proliferation, migration, and matrix synthesis. Once the injury has resolved, vSMC redifferentiation ensues with restored contractile response, and low proliferative and migratory activity. Chronic injury can lead to vSMC hyperplasia or excessive proliferation for instance during neointima formation or atherosclerosis.[116] Both conditions correlate with low RGS5 expression.[30,59] PDGF and the prototypic RTK PDGFR induce vSMC migration and switching to a synthetic phenotype.[27,117] Surprisingly, RGS5 attenuates PDGF-induced ERK phosphorylation suggesting a regulatory role in receptor RTK-GPCR cross talk.[27,118] As shown by Gunaje *et al.*, vascular injury-induced PDGFR activation downregulates RGS5 which in turn releases the brake on GPCR signaling and permits vSMC migration/proliferation and—if unresolved—hypertrophy via MEK/ERK, pAKT, and pJnk signaling.[100] While this is an attractive concept which links RTK and GPCR involving RGS5, it remains to be shown how PDGFR signaling regulates RGS5 levels.

RGS5 deficiency exacerbates cardiac and liver fibrosis implicating RGS5 in TGFβ-RTK signaling at the cross road between MEK/ERK and Smad activation.[23,91] Interestingly, there is precedence for another RGS family member, the G protein-coupled receptor kinase-2 (GRK2), to integrate TGFβ-ERK signal transduction cascades. GRK2 expression is increased by TGFβ in vSMC which in turn antagonizes AngII-induced ERK phosphorylation, vSMC proliferation, and migration.[119] Thus, the GPCR-regulator GRK2 provides negative feedback to terminate TGFβ-induced Smad signaling which is remarkably similar to RGS5's role in cardiac fibrosis. So far, direct regulation of RGS5 by TGFβ has only been reported in a hepatic stellate cell line *in vitro*.[91]

5.2 Does RGS5 Balance Signaling Between G Proteins Subunits?

There is strong evidence for direct RGS5 interaction with Gαi and Gαq proteins[17] consistent with a role of both subunits in GPCR-activated vessel contraction and the MAP kinase pathway.[23] As yet, RGS5 is not known to act as GAP for Gαs subunits. However, under chronic β adrenergic activation of the PKA-cAMP axis in aSMC, RGS5 levels decrease concomitant with sensitization of contractile pathways suggesting a role of RGS5 in the switch from Gαs to Gαq signaling[55] (Fig. 4A). Conversely, in the heart, coupling of β2-adrenoceptors to Gi/o proteins may be blunted by upregulation of RGS5 concomitant with increased Gαs signaling under chronic conditions.[36]

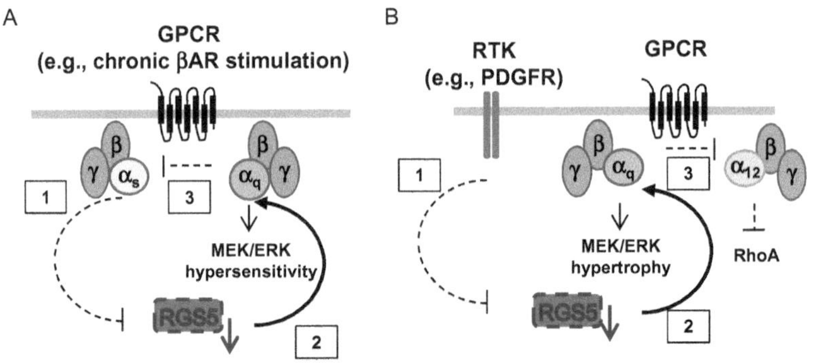

Figure 4 Hypothetical role of RGS5 as modulator between different G protein subtypes and receptor prototypes. (A) Chronic stimulation of Gαs-conjugated β adrenergic receptor in airway smooth muscle cells results in downregulation of RGS5 and increased Gαq-mediated signaling which leads to aSMC hyper-responsiveness.[55] (B) PDGFR activation following vascular injury downregulates RGS5 which in turn enhances GPCR signaling via Gαq. The net outcome of sustained injury is increased vSMC migration/ proliferation and hypertrophy.[100] Increase in Gαq signaling may decrease Gα12/13 and RhoA activation.[48] (See the color plate.)

RGS5 GAP activity for Gα12/13G proteins has not been shown although a potential functional link between Gα12/13 and RGS5 has recently been reported in a model of arteriogenesis.[48] In this study, loss of RGS5 hyperactivates Gαq signaling which in turn impairs Gα12/13-mediated RhoA stimulation, indicating that RGS5 levels may shift the balance between both signal transduction pathways. Such an antagonistic role of RGS5 implies divergent roles of Gαq and Gα12/13 signaling in blood vessels. Interestingly, Gαq and Gα12/13 have synergistic functions in the regulation of vessel contraction and blood pressure homeostasis.[120] Nevertheless, Gαq and Gα12/13 also act antagonistically to regulate vSMC phenotype switching during vascular remodeling in atherosclerosis or vascular injury.[96] In vSMC-specific knockout studies of Gαq or Gα12/13 subunits, Althoff *et al.* demonstrated that predominant Gαq signaling via ERK results in excessive vascular remodeling whereas enhanced Gα12/13-mediated signaling through RhoA promotes vSMC differentiation. Specifically, in the absence of Gα12/13 in vSMC, enhanced Gαq signaling activates ERK and ternary complex factors (TCF) such as Elk which bind to serum response factor (SRF) and initiate transcription of SRF response genes involved in vascular proliferation. Simultaneously, absence of RhoA activation abolishes expression/nuclear translocation of an alternative set of SRF binding factors such as myocardin and myocardin-related transcription factors (MRTF) which drive differentiation;

thus, the net outcome of increased Gαq versus Gα12/13 activity is vSMC dedifferentiation.[96] So far, it remains unknown how Gαq and Gα12/13 signaling pathways are antagonistically activated and fine tuned. This may involve differential receptor coupling, or, as suggested by Althoff *et al.*, RGS proteins which could regulate the balance between MEK/ERK and RhoA signaling in response to the same receptor.[96] Hypothetically then, if RGS5 is downregulated by PDGF—which so far has only been shown *in vitro*[100]—this would lead to Gαq-biased signaling and vSMC dedifferentiation; once vascular injury resolves and PDGF levels subside, RGS5 upregulation would then shift signaling toward Gα12/13 and stimulate redifferentiation (Fig. 4B). While this is an interesting model, it is also likely to be oversimplified. For instance, it remains unclear whether inhibition of one pathway results in stimulation of the other "by default," and indeed how RGS5 levels are regulated during vascular remodeling *in vivo*. Moreover, vSMC-specific Gα12/13 deficiency and enhanced Gαq activity increase vSMC proliferation and neointima hyperplasia after carotid artery ligation[96]; the opposite has been observed in RGS5-deficient mice where neointima formation is attenuated after artery ligation compared to wild-type animals.[48] Since loss of RGS5 exacerbates Gαq and MEK/ERK signaling, the latter result appears counterintuitive. Thus, while RGS5 clearly regulates both, MEK/ERK and RhoA signaling, its potential endogenous role in fine tuning the response to vascular injury still remains to be elucidated.

6. THE CARDIOVASCULAR RGS FAMILY: IT'S A FINE BALANCE

RGS proteins do not simply switch signaling cascades on or off. Instead, they modulate and coordinate incoming signals in a highly context-dependent manner. In fact, the endogenous cardiovascular roles of RGS2, RGS4, and RGS5 exemplify the exquisite control RGS molecules exert and the fine balance between gain and loss of function. For instance, too much RGS2 protein is implicated in a human pathological condition of abnormal vascular tone regulation, the so-called Bartter/Gitelman syndrome, where patients have subdued AngII signaling on a background of high circulating AngII levels.[121] Consistently, loss of RGS2 in knockout animals results in hypertension,[46] and rare RGS2-destabilizing mutations have been identified in hypertensive human individuals.[40,122] Moreover, RGS2 is required for early myocardial compensation to pressure overload since RGS2$^{-/-}$ mice fail to counteract maladaptive cardiac hypertrophy; loss of RGS2 also exacerbates atrial arrhythmias.[34,123,124] Even though murine knockout studies indicate a non-redundant role for RGS5, it shares many features with RGS2 and RGS4. For instance, as a prominent regulator of AngII signaling, loss of RGS5 induces hypertension[38] consistent with population data which implicate RGS5 polymorphisms in essential hypertension.[42–44] Similarly, cardiac hypertrophy and arrhythmias are prominent in RGS5-deficient animals, suggesting RGS5 protects from heart failure.[23,49,50] In contrast, upregulation of RGS5 in tumor vessels enhances angiogenesis and cancer aggressiveness.[22,84] Interestingly, the capacity to regulate angiogenesis and intratumoral barrier function as well as being regulated by hypoxia seem to be unique RGS5 features.[99] Endogenous RGS4—in contrast to RGS5—is predominantly expressed in coronary arteries and the sinoatrial node (SAN) and has been implicated in the parasympathetic control of heart rate.[125,126] However, most recently, RGS6 has been identified as a regulator of a potentially dominant endogenous pacemaker function.[127] Genetic deletion of RGS6 increases response to M2 muscarinic receptor stimulation and parasympathetic activity of the heart.[128,129] Unexpectedly, mice deficient for both, RGS4 and 6, are partially protected and the existence of yet another R7 RGS family member was proposed which may functionally replace RGS6 in the absence of RGS4.[127] These latest studies demonstrate a complex regulatory circuit in a single specialized compartment (SAN) and potential redundancy among RGS molecules which was only uncovered in double knockout studies. Unfortunately, generation of more than one gene knockout in particular for the B/R4 RGS family remains technologically challenging since most genes cluster on chromosome 1 (RGS1, RGS2, RGS4, RGS5, RGS8, RGS13, RGS18, RGS16).

RGS proteins are subject to rapid proteasomal degradation and turn over. Thus, pharmacological protein stabilization has been proposed as a potential therapeutic approach for instance to increase RGS2 in hypertension, or RGS4 in metastatic breast cancer.[130,131] However, much as their physiological role is context dependent, designing balanced strategies to modulate their expression and function in pathophysiology is even more critical. For example, too much RGS5 correlates with poor cancer prognosis but too little causes hypertension; hence, blocking tumor angiogenesis may result in high blood pressure and reduced capability for collateral arterial growth during ischemia.[22,38,48] Thus, it is essential to consider complexities of shared pathways when exploring the potential clinical use of RGS inhibitors and agonists.

7. CHALLENGES IN THE FIELD

The last decade has witnessed enormous progress in the understanding of RGS protein function in laboratory animals and humans. Given the prevalence of GPCR signaling in most biological processes, it is not surprising that downstream modulators of GPCRs such as RGS proteins are also interesting therapeutic targets. However, the field is plagued with inconsistent and sometimes diametrically opposite findings depending on model systems and applied technologies[19,38] (Table 1). Often RGS proteins are studied out of context in isolated cells or tissue with little endogenous expression.[132] The dynamic regulation of RGS5 has almost exclusively been studied on a transcriptional level when in fact RGS5 protein stability is highly controlled, and its subcellular location and activity may heavily depend on post-translational modifications.[133] Anti-RGS5 antibodies are generally of poor quality and specificity; nevertheless, they are frequently used without appropriate knockdown or knockout controls,[78,79] or major conclusions are based on semi-quantitative assays such as immunohistochemistry.[48] Moreover, endogenous RGS5 levels in organs are low and thus difficult to detect without enrichment strategies.[57] Although RGS5 is a prominent cardiovascular protein, many features among B/R4 family members are shared, and much of its specificity for particular receptors and G proteins under physiological conditions remains undefined.[6,18] Therefore, the challenge ahead is to gain insights into endogenous RGS5 protein function and regulation in a spatiotemporal context *in vivo* which requires, among others, sophisticated real-time imaging technology. In particular, more information about intracellular trafficking between plasma membrane, cytoplasm, or even nucleus, as well as association with non-Gα proteins such as 14-3-3 scaffold proteins or clathrin adaptor proteins is needed for a better conceptual understanding of RGS5 biology.[134–136]

So far, development of specific inhibitors or agonists for RGS molecules, which would greatly assist with this task, has been slow and clearly focused on RGS4. Small molecule inhibitors which target RGS4 are designed to disrupt RGS4-Gα interactions and provide proof of concept that specific targeting is feasible.[137] As yet however, no synthetic RGS5 agonist or antagonist is available. Nevertheless, recent insights into endogenous RGS5 protein and mRNA regulation may open new options for specific modulation, for instance by activating gene transcription via specific promoter methylation,[28] binding of PPAR agonists or antagonists,[102,106] or increasing protein stability by blocking N-end rule protein degradation.[97] The role of RGS5 in vascular pathologies has just emerged but to date few molecules have been described with such a key function in vascular remodeling. In particular, RGS5's role in receptor desensitization, receptor cross talk, and G protein antagonism requires deeper mechanistic insights but will most certainly open new and expanded opportunities for therapeutic application.

ACKNOWLEDGMENTS

The author acknowledges financial support from grants and fellowships from the Western Australian Cancer Council, the Australian National Health and Medical Research Council (NHMRC), the Royal Perth Hospital Medical Research Foundation, and the Harry Perkins Institute of Medical Research.

REFERENCES

1. De Vries L, Zheng B, Fischer T, Elenko E, Farquhar MG. The regulator of G protein signaling family. *Annu Rev Pharmacol Toxicol.* 2000;40:235–271.
2. Sjogren B, Neubig RR. Thinking outside of the "RGS box": new approaches to therapeutic targeting of regulators of G protein signaling. *Mol Pharmacol.* 2010;78(4): 550–557.
3. Sethakorn N, Yau DM, Dulin NO. Non-canonical functions of RGS proteins. *Cell Signal.* 2010;22(9):1274–1281.
4. Cho H, Harrison K, Schwartz O, Kehrl JH. The aorta and heart differentially express RGS (regulators of G-protein signalling) proteins that selectively regulate sphingosine 1-phosphate, angiotensin II and endothelin-1 signalling. *Biochem J.* 2003;371(pt 3): 973–980.
5. Kardestuncer T, Wu H, Lim AL, Neer EJ. Cardiac myocytes express mRNA for ten RGS proteins: changes in RGS mRNA expression in ventricular myocytes and cultured atria. *FEBS Lett.* 1998;438(3):285–288.
6. Wang Q, Liu M, Mullah B, Siderovski DP, Neubig RR. Receptor-selective effects of endogenous RGS3 and RGS5 to regulate mitogen-activated protein kinase activation in rat vascular smooth muscle cells. *J Biol Chem.* 2002;277(28):24949–24958.
7. Harris IS, Treskov I, Rowley MW, et al. G-protein signaling participates in the development of diabetic cardiomyopathy. *Diabetes.* 2004;53(12):3082–3090.
8. Waterson RE, Thompson CG, Mabe NW, et al. Galpha(i2)-mediated protection from ischaemic injury is modulated by endogenous RGS proteins in the mouse heart. *Cardiovasc Res.* 2011;91(1):45–52.
9. Signarvic RS, Cierniewska A, Stalker TJ, et al. RGS/Gi2alpha interactions modulate platelet accumulation and thrombus formation at sites of vascular injury. *Blood.* 2010;116(26):6092–6100.
10. Louwette S, Van Geet C, Freson K. Regulators of G protein signaling: role in hematopoiesis, megakaryopoiesis and platelet function. *J Thromb Haemost.* 2012;10(11): 2215–2222.
11. Stewart A, Huang J, Fisher RA. RGS proteins in heart: brakes on the vagus. *Front Physiol.* 2012;3:95.
12. Manzur M, Ganss R. Regulator of G protein signaling 5: a new player in vascular remodeling. *Trends Cardiovasc Med.* 2009;19(1):26–30.

13. Gu S, Cifelli C, Wang S, Heximer SP. RGS proteins: identifying new GAPs in the understanding of blood pressure regulation and cardiovascular function. *Clin Sci (Lond)*. 2009;116(5):391–399.
14. Zhang P, Mende U. Regulators of G-protein signaling in the heart and their potential as therapeutic targets. *Circ Res*. 2011;109(3):320–333.
15. Chen C, Zheng B, Han J, Lin SC. Characterization of a novel mammalian RGS protein that binds to Galpha proteins and inhibits pheromone signaling in yeast. *J Biol Chem*. 1997;272(13):8679–8685.
16. Seki N, Hattori A, Sugano S, et al. Isolation, tissue expression, and chromosomal assignment of a novel human gene which encodes a protein with RING finger motif. *J Hum Genet*. 1998;43(4):272–274.
17. Zhou J, Moroi K, Nishiyama M, et al. Characterization of RGS5 in regulation of G protein-coupled receptor signaling. *Life Sci*. 2001;68(13):1457–1469.
18. Hao J, Michalek C, Zhang W, Zhu M, Xu X, Mende U. Regulation of cardiomyocyte signaling by RGS proteins: differential selectivity towards G proteins and susceptibility to regulation. *J Mol Cell Cardiol*. 2006;41(1):51–61.
19. Cho H, Park C, Hwang IY, et al. Rgs5 targeting leads to chronic low blood pressure and a lean body habitus. *Mol Cell Biol*. 2008;28(8):2590–2597.
20. Ketsawatsomkron P, Lorca RA, Keen HL, et al. PPARgamma regulates resistance vessel tone through a mechanism involving RGS5-mediated control of protein kinase C and BKCa channel activity. *Circ Res*. 2012;111(11):1446–1458.
21. Nisancioglu MH, Mahoney Jr WM, Kimmel DD, Schwartz SM, Betsholtz C, Genove G. Generation and characterization of rgs5 mutant mice. *Mol Cell Biol*. 2008;28(7):2324–2331.
22. Hamzah J, Jugold M, Kiessling F, et al. Vascular normalization in Rgs5-deficient tumours promotes immune destruction. *Nature*. 2008;453(7193):410–414.
23. Li H, He C, Feng J, et al. Regulator of G protein signaling 5 protects against cardiac hypertrophy and fibrosis during biomechanical stress of pressure overload. *Proc Natl Acad Sci USA*. 2010;107(31):13818–13823.
24. Adams LD, Geary RL, McManus B, Schwartz SM. A comparison of aorta and vena cava medial message expression by cDNA array analysis identifies a set of 68 consistently differentially expressed genes, all in aortic media. *Circ Res*. 2000;87(7):623–631.
25. Li J, Adams LD, Wang X, et al. Regulator of G protein signaling 5 marks peripheral arterial smooth muscle cells and is downregulated in atherosclerotic plaque. *J Vasc Surg*. 2004;40(3):519–528.
26. Bondjers C, Kalen M, Hellstrom M, et al. Transcription profiling of platelet-derived growth factor-B-deficient mouse embryos identifies RGS5 as a novel marker for pericytes and vascular smooth muscle cells. *Am J Pathol*. 2003;162(3):721–729.
27. Cho H, Kozasa T, Bondjers C, Betsholtz C, Kehrl JH. Pericyte-specific expression of Rgs5: implications for PDGF and EDG receptor signaling during vascular maturation. *FASEB J*. 2003;17(3):440–442.
28. Zhang H, Gu S, Al-Sabeq B, et al. Origin-specific epigenetic program correlates with vascular bed-specific differences in Rgs5 expression. *FASEB J*. 2012;26(1):181–191.
29. Kirsch T, Wellner M, Luft FC, Haller H, Lippoldt A. Altered gene expression in cerebral capillaries of stroke-prone spontaneously hypertensive rats. *Brain Res*. 2001;910(1–2):106–115.
30. Geary RL, Wong JM, Rossini A, Schwartz SM, Adams LD. Expression profiling identifies 147 genes contributing to a unique primate neointimal smooth muscle cell phenotype. *Arterioscler Thromb Vasc Biol*. 2002;22(12):2010–2016.
31. Berger M, Bergers G, Arnold B, Hammerling GJ, Ganss R. Regulator of G-protein signaling-5 induction in pericytes coincides with active vessel remodeling during neovascularization. *Blood*. 2005;105(3):1094–1101.

32. Grayson TH, Ohms SJ, Brackenbury TD, et al. Vascular microarray profiling in two models of hypertension identifies caveolin-1, Rgs2 and Rgs5 as antihypertensive targets. *BMC Genomics*. 2007;8:404.
33. Riddle EL, Schwartzman RA, Bond M, Insel PA. Multi-tasking RGS proteins in the heart: the next therapeutic target? *Circ Res*. 2005;96(4):401–411.
34. Takimoto E, Koitabashi N, Hsu S, et al. Regulator of G protein signaling 2 mediates cardiac compensation to pressure overload and antihypertrophic effects of PDE5 inhibition in mice. *J Clin Invest*. 2009;119(2):408–420.
35. Tokudome T, Kishimoto I, Horio T, et al. Regulator of G-protein signaling subtype 4 mediates antihypertrophic effect of locally secreted natriuretic peptides in the heart. *Circulation*. 2008;117(18):2329–2339.
36. Jean-Baptiste G, Li X, Yang Z, et al. Beta adrenergic receptor-mediated atrial specific up-regulation of RGS5. *Life Sci*. 2005;76(13):1533–1545.
37. Hayashida T, Decaestecker M, Schnaper HW. Cross-talk between ERK MAP kinase and Smad signaling pathways enhances TGF-beta-dependent responses in human mesangial cells. *FASEB J*. 2003;17(11):1576–1578.
38. Holobotovskyy V, Manzur M, Tare M, et al. Regulator of G-protein signaling 5 controls blood pressure homeostasis and vessel wall remodeling. *Circ Res*. 2013;112(5):781–791.
39. Wang X, Adams LD, Pabon LM, et al. RGS5, RGS4, and RGS2 expression and aortic contractibility are dynamically co-regulated during aortic banding-induced hypertrophy. *J Mol Cell Cardiol*. 2008;44(3):539–550.
40. Yang J, Kamide K, Kokubo Y, et al. Genetic variations of regulator of G-protein signaling 2 in hypertensive patients and in the general population. *J Hypertens*. 2005;23(8):1497–1505.
41. Riddle EL, Rana BK, Murthy KK, et al. Polymorphisms and haplotypes of the regulator of G protein signaling-2 gene in normotensives and hypertensives. *Hypertension*. 2006;47(3):415–420.
42. Chang YP, Liu X, Kim JD, et al. Multiple genes for essential-hypertension susceptibility on chromosome 1q. *Am J Hum Genet*. 2007;80(2):253–264.
43. Faruque MU, Chen G, Doumatey A, et al. Association of ATP1B1, RGS5 and SELE polymorphisms with hypertension and blood pressure in African-Americans. *J Hypertens*. 2011;29(10):1906–1912.
44. Xiao B, Zhang Y, Niu WQ, Gao PJ, Zhu DL. Haplotype-based association of regulator of G-protein signaling 5 gene polymorphisms with essential hypertension and metabolic parameters in Chinese. *Clin Chem Lab Med*. 2009;47(12):1483–1488.
45. Tang M, Wang G, Lu P, et al. Regulator of G-protein signaling-2 mediates vascular smooth muscle relaxation and blood pressure. *Nat Med*. 2003;9(12):1506–1512.
46. Heximer SP, Knutsen RH, Sun X, et al. Hypertension and prolonged vasoconstrictor signaling in RGS2-deficient mice. *J Clin Invest*. 2003;111(8):1259.
47. Hercule HC, Tank J, Plehm R, et al. Regulator of G protein signalling 2 ameliorates angiotensin II-induced hypertension in mice. *Exp Physiol*. 2007;92(6):1014–1022.
48. Arnold C, Feldner A, Pfisterer L, et al. RGS5 promotes arterial growth during arteriogenesis. *EMBO Mol Med*. 2014;6(8):1075–1089.
49. Qin M, Huang H, Wang T, et al. Absence of Rgs5 prolongs cardiac repolarization and predisposes to ventricular tachyarrhythmia in mice. *J Mol Cell Cardiol*. 2012;53(6): 880–890.
50. Qin M, Huang H, Wang T, et al. Atrial tachyarrhythmia in rgs5-null mice. *PLoS One*. 2012;7(11):e46856.
51. Gross V, Tank J, Obst M, et al. Autonomic nervous system and blood pressure regulation in RGS2-deficient mice. *Am J Physiol Regul Integr Comp Physiol*. 2005;288(5): R1134–R1142.

52. Prakash YS. Airway smooth muscle in airway reactivity and remodeling: what have we learned? *Am J Physiol Lung Cell Mol Physiol*. 2013;305(12):L912–L933.
53. Finney PA, Belvisi MG, Donnelly LE, et al. Albuterol-induced downregulation of Gsalpha accounts for pulmonary beta(2)-adrenoceptor desensitization in vivo. *J Clin Invest*. 2000;106(1):125–135.
54. Mak JC, Roffel AF, Katsunuma T, Elzinga CR, Zaagsma J, Barnes PJ. Up-regulation of airway smooth muscle histamine H(1) receptor mRNA, protein, and function by beta(2)-adrenoceptor activation. *Mol Pharmacol*. 2000;57(5):857–864.
55. Yang Z, Cooper PR, Damera G, et al. Beta-agonist-associated reduction in RGS5 expression promotes airway smooth muscle hyper-responsiveness. *J Biol Chem*. 2011;286(13):11444–11455.
56. Labuda M, Laberge S, Briere J, Berube D, Krajinovic M. RGS5 gene and therapeutic response to short acting bronchodilators in paediatric asthma patients. *Pediatr Pulmonol*. 2013;48(10):970–975.
57. Yang Z, Balenga N, Cooper PR, et al. Regulator of G-protein signaling-5 inhibits bronchial smooth muscle contraction in severe asthma. *Am J Respir Cell Mol Biol*. 2012;46(6):823–832.
58. Adams LD, Geary RL, Li J, Rossini A, Schwartz SM. Expression profiling identifies smooth muscle cell diversity within human intima and plaque fibrous cap: loss of RGS5 distinguishes the cap. *Arterioscler Thromb Vasc Biol*. 2006;26(2):319–325.
59. Takata Y, Liu J, Yin F, et al. PPARdelta-mediated antiinflammatory mechanisms inhibit angiotensin II-accelerated atherosclerosis. *Proc Natl Acad Sci USA*. 2008;105(11):4277–4282.
60. Barish GD, Atkins AR, Downes M, et al. PPARdelta regulates multiple proinflammatory pathways to suppress atherosclerosis. *Proc Natl Acad Sci USA*. 2008;105(11):4271–4276.
61. Liu SL, Li YH, Shi GY, Jiang MJ, Chang JH, Wu HL. The effect of statin on the aortic gene expression profiling. *Int J Cardiol*. 2007;114(1):71–77.
62. Deng W, Wang X, Xiao J, et al. Loss of regulator of G protein signaling 5 exacerbates obesity, hepatic steatosis, inflammation and insulin resistance. *PLoS One*. 2012;7(1):e30256.
63. Dunn J, Qiu H, Kim S, et al. Flow-dependent epigenetic DNA methylation regulates endothelial gene expression and atherosclerosis. *J Clin Invest*. 2014;124(7):3187–3199.
64. Manome Y, Saeki N, Yoshinaga H, Watanabe M, Mizuno S. A culture device demonstrates that hydrostatic pressure increases mRNA of RGS5 in neuroblastoma and CHC1-L in lymphocytic cells. *Cells Tissues Organs*. 2003;174(4):155–161.
65. Bergers G, Song S. The role of pericytes in blood-vessel formation and maintenance. *Neuro Oncol*. 2005;7(4):452–464.
66. Morikawa S, Baluk P, Kaidoh T, Haskell A, Jain RK, McDonald DM. Abnormalities in pericytes on blood vessels and endothelial sprouts in tumors. *Am J Pathol*. 2002;160(3):985–1000.
67. Song S, Ewald AJ, Stallcup W, Werb Z, Bergers G. PDGFRbeta+ perivascular progenitor cells in tumours regulate pericyte differentiation and vascular survival. *Nat Cell Biol*. 2005;7(9):870–879.
68. Ryschich E, Schmidt J, Hammerling GJ, Klar E, Ganss R. Transformation of the microvascular system during multistage tumorigenesis. *Int J Cancer*. 2002;97(6):719–725.
69. Bergers G, Benjamin LE. Tumorigenesis and the angiogenic switch. *Nat Rev Cancer*. 2003;3(6):401–410.
70. Mitchell TS, Bradley J, Robinson GS, Shima DT, Ng YS. RGS5 expression is a quantitative measure of pericyte coverage of blood vessels. *Angiogenesis*. 2008;11(2):141–151.

71. Manzur M, Hamzah J, Ganss R. Modulation of the "blood-tumor" barrier improves immunotherapy. *Cell Cycle*. 2008;7(16):2452–2455.
72. Armulik A, Genove G, Mae M, et al. Pericytes regulate the blood–brain barrier. *Nature*. 2010;468(7323):557–561.
73. Proebstl D, Voisin MB, Woodfin A, et al. Pericytes support neutrophil subendothelial cell crawling and breaching of venular walls in vivo. *J Exp Med*. 2012;209(6): 1219–1234.
74. Stark K, Eckart A, Haidari S, et al. Capillary and arteriolar pericytes attract innate leukocytes exiting through venules and 'instruct' them with pattern-recognition and motility programs. *Nat Immunol*. 2013;14(1):41–51.
75. Bose A, Barik S, Banerjee S, et al. Tumor-derived vascular pericytes anergize Th cells. *J Immunol*. 2013;191(2):971–981.
76. Chen X, Higgins J, Cheung ST, et al. Novel endothelial cell markers in hepatocellular carcinoma. *Mod Pathol*. 2004;17(10):1198–1210.
77. Furuya M, Nishiyama M, Kimura S, et al. Expression of regulator of G protein signalling protein 5 (RGS5) in the tumour vasculature of human renal cell carcinoma. *J Pathol*. 2004;203(1):551–558.
78. Silini A, Ghilardi C, Figini S, et al. Regulator of G-protein signaling 5 (RGS5) protein: a novel marker of cancer vasculature elicited and sustained by the tumor's proangiogenic microenvironment. *Cell Mol Life Sci*. 2012;69(7):1167–1178.
79. Wang JH, Huang WS, Hu CR, Guan XX, Zhou HB, Chen LB. Relationship between RGS5 expression and differentiation and angiogenesis of gastric carcinoma. *World J Gastroenterol*. 2010;16(44):5642–5646.
80. Stiles JM, Rowntree RK, Amaya C, et al. Gene expression analysis reveals marked differences in the transcriptome of infantile hemangioma endothelial cells compared to normal dermal microvascular endothelial cells. *Vasc Cell*. 2013;5(1):6.
81. Boss CN, Grunebach F, Brauer K, et al. Identification and characterization of T-cell epitopes deduced from RGS5, a novel broadly expressed tumor antigen. *Clin Cancer Res*. 2007;13(11):3347–3355.
82. Koh J, Dar M, Untch BR, et al. Regulator of G protein signaling 5 is highly expressed in parathyroid tumors and inhibits signaling by the calcium-sensing receptor. *Mol Endocrinol*. 2011;25(5):867–876.
83. Huang G, Song H, Wang R, Han X, Chen L. The relationship between RGS5 expression and cancer differentiation and metastasis in non-small cell lung cancer. *J Surg Oncol*. 2012;105(4):420–424.
84. Hu M, Chen X, Zhang J, et al. Over-expression of regulator of G protein signaling 5 promotes tumor metastasis by inducing epithelial-mesenchymal transition in hepatocellular carcinoma cells. *J Surg Oncol*. 2013;108(3):192–196.
85. Yao M, Huang Y, Shioi K, et al. A three-gene expression signature model to predict clinical outcome of clear cell renal carcinoma. *Int J Cancer*. 2008;123(5): 1126–1132.
86. Volz NB, Stintzing S, Zhang W, et al. Genes involved in pericyte-driven tumor maturation predict treatment benefit of first-line FOLFIRI plus bevacizumab in patients with metastatic colorectal cancer. *Pharmacogenomics J*. 2015;15(1):69–76.
87. Lovschall H, Mitsiadis TA, Poulsen K, Jensen KH, Kjeldsen AL. Coexpression of Notch3 and Rgs5 in the pericyte-vascular smooth muscle cell axis in response to pulp injury. *Int J Dev Biol*. 2007;51(8):715–721.
88. Crisan M, Yap S, Casteilla L, et al. A perivascular origin for mesenchymal stem cells in multiple human organs. *Cell Stem Cell*. 2008;3(3):301–313.
89. Paquet-Fifield S, Schluter H, Li A, et al. A role for pericytes as microenvironmental regulators of human skin tissue regeneration. *J Clin Invest*. 2009;119(9): 2795–2806.

90. Fleming JN, Nash RA, McLeod DO, et al. Capillary regeneration in scleroderma: stem cell therapy reverses phenotype? *PLoS One*. 2008;3(1):e1452.
91. Bahrami AJ, Gunaje JJ, Hayes BJ, et al. Regulator of g-protein signaling-5 is a marker of hepatic stellate cells and expression mediates response to liver injury. *PLoS One*. 2014;9(10):e108505.
92. Mills SJ, Cowin AJ, Kaur P. Pericytes, mesenchymal stem cells and the wound healing process. *Cells*. 2013;2(3):621–634.
93. Troidl K, Schaper W. Arteriogenesis versus angiogenesis in peripheral artery disease. *Diabetes Metab Res Rev*. 2012;28(suppl 1):27–29.
94. Heil M, Schaper W. Pathophysiology of collateral development. *Coron Artery Dis*. 2004;15(7):373–378.
95. Wuertz CM, Lorincz A, Vettel C, Thomas MA, Wieland T, Lutz S. p63RhoGEF—a key mediator of angiotensin II-dependent signaling and processes in vascular smooth muscle cells. *FASEB J*. 2010;24(12):4865–4876.
96. Althoff TF, Albarran Juarez J, Troidl K, et al. Procontractile G protein-mediated signaling pathways antagonistically regulate smooth muscle differentiation in vascular remodeling. *J Exp Med*. 2012;209(12):2277–2290.
97. Lee MJ, Tasaki T, Moroi K, et al. RGS4 and RGS5 are in vivo substrates of the N-end rule pathway. *Proc Natl Acad Sci USA*. 2005;102(42):15030–15035.
98. Hu RG, Sheng J, Qi X, Xu Z, Takahashi TT, Varshavsky A. The N-end rule pathway as a nitric oxide sensor controlling the levels of multiple regulators. *Nature*. 2005;437(7061):981–986.
99. Jin Y, An X, Ye Z, Cully B, Wu J, Li J. RGS5, a hypoxia-inducible apoptotic stimulator in endothelial cells. *J Biol Chem*. 2009;284(35):23436–23443.
100. Gunaje JJ, Bahrami AJ, Schwartz SM, Daum G, Mahoney Jr WM. PDGF-dependent regulation of regulator of G protein signaling-5 expression and vascular smooth muscle cell functionality. *Am J Physiol Cell Physiol*. 2011;301(2):C478–C489.
101. Nunn C, Zou MX, Sobiesiak AJ, Roy AA, Kirshenbaum LA, Chidiac P. RGS2 inhibits beta-adrenergic receptor-induced cardiomyocyte hypertrophy. *Cell Signal*. 2010;22(8):1231–1239.
102. Zarzuelo MJ, Jimenez R, Galindo P, et al. Antihypertensive effects of peroxisome proliferator-activated receptor-beta activation in spontaneously hypertensive rats. *Hypertension*. 2011;58(4):733–743.
103. Zarzuelo MJ, Gomez-Guzman M, Jimenez R, et al. Effects of peroxisome proliferator-activated receptor-beta activation in endothelin-dependent hypertension. *Cardiovasc Res*. 2013;99(4):622–631.
104. Barish GD, Narkar VA, Evans RM. PPAR delta: a dagger in the heart of the metabolic syndrome. *J Clin Invest*. 2006;116(3):590–597.
105. Ketsawatsomkron P, Pelham CJ, Groh S, Keen HL, Faraci FM, Sigmund CD. Does peroxisome proliferator-activated receptor-gamma (PPAR gamma) protect from hypertension directly through effects in the vasculature? *J Biol Chem*. 2010;285(13):9311–9316.
106. Ketsawatsomkron P, Lorca RA, Keen HL, et al. PPARgamma regulates resistance vessel tone through a mechanism involving RGS5-mediated control of PKC and BKCa channel activity. *Circ Res*. 2012;111(11):1446–1458.
107. Chen C, Seow KT, Guo K, Yaw LP, Lin SC. The membrane association domain of RGS16 contains unique amphipathic features that are conserved in RGS4 and RGS5. *J Biol Chem*. 1999;274(28):19799–19806.
108. Bernstein LS, Grillo AA, Loranger SS, Linder ME. RGS4 binds to membranes through an amphipathic alpha -helix. *J Biol Chem*. 2000;275(24):18520–18526.
109. Kwon YT, Kashina AS, Davydov IV, et al. An essential role of N-terminal arginylation in cardiovascular development. *Science*. 2002;297(5578):96–99.

110. Lee MJ, Kim DE, Zakrzewska A, et al. Characterization of arginylation branch of N-end rule pathway in G-protein-mediated proliferation and signaling of cardiomyocytes. *J Biol Chem*. 2012;287(28):24043–24052.
111. Jaba IM, Zhuang ZW, Li N, et al. NO triggers RGS4 degradation to coordinate angiogenesis and cardiomyocyte growth. *J Clin Invest*. 2013;123(4):1718–1731.
112. Wang J, Xie Y, Wolff DW, Abel PW, Tu Y. DHHC protein-dependent palmitoylation protects regulator of G-protein signaling 4 from proteasome degradation. *FEBS Lett*. 2010;584(22):4570–4574.
113. Brower CS, Varshavsky A. Ablation of arginylation in the mouse N-end rule pathway: loss of fat, higher metabolic rate, damaged spermatogenesis, and neurological perturbations. *PLoS One*. 2009;4(11):e7757.
114. Moroi K, Nishiyama M, Kawabata S, Ichiba H, Yajima T, Kimura S. Phosphorylation of Ser166 in RGS5 by protein kinase C causes loss of RGS function. *Life Sci*. 2007;81(1):40–50.
115. Owens GK, Kumar MS, Wamhoff BR. Molecular regulation of vascular smooth muscle cell differentiation in development and disease. *Physiol Rev*. 2004;84(3):767–801.
116. Davis-Dusenbery BN, Wu C, Hata A. Micromanaging vascular smooth muscle cell differentiation and phenotypic modulation. *Arterioscler Thromb Vasc Biol*. 2011;31(11): 2370–2377.
117. Deaton RA, Gan Q, Owens GK. Sp1-dependent activation of KLF4 is required for PDGF-BB-induced phenotypic modulation of smooth muscle. *Am J Physiol Heart Circ Physiol*. 2009;296(4):H1027–H1037.
118. Wetzker R, Bohmer FD. Transactivation joins multiple tracks to the ERK/MAPK cascade. *Nat Rev Mol Cell Biol*. 2003;4(8):651–657.
119. Guo J, Chen H, Ho J, et al. TGFbeta-induced GRK2 expression attenuates AngII-regulated vascular smooth muscle cell proliferation and migration. *Cell Signal*. 2009;21(6):899–905.
120. Wirth A, Benyo Z, Lukasova M, et al. G12-G13-LARG-mediated signaling in vascular smooth muscle is required for salt-induced hypertension. *Nat Med*. 2008;14(1):64–68.
121. Calo LA, Pagnin E, Davis PA, et al. Increased expression of regulator of G protein signaling-2 (RGS-2) in Bartter's/Gitelman's syndrome. A role in the control of vascular tone and implication for hypertension. *J Clin Endocrinol Metab*. 2004;89(8):4153–4157.
122. Bodenstein J, Sunahara RK, Neubig RR. N-terminal residues control proteasomal degradation of RGS2, RGS4, and RGS5 in human embryonic kidney 293 cells. *Mol Pharmacol*. 2007;71(4):1040–1050.
123. Zhang W, Anger T, Su J, et al. Selective loss of fine tuning of Gq/11 signaling by RGS2 protein exacerbates cardiomyocyte hypertrophy. *J Biol Chem*. 2006;281(9):5811–5820.
124. Tuomi JM, Chidiac P, Jones DL. Evidence for enhanced M3 muscarinic receptor function and sensitivity to atrial arrhythmia in the RGS2-deficient mouse. *Am J Physiol Heart Circ Physiol*. 2010;298(2):H554–H561.
125. Grillet N, Pattyn A, Contet C, Kieffer BL, Goridis C, Brunet JF. Generation and characterization of Rgs4 mutant mice. *Mol Cell Biol*. 2005;25(10):4221–4228.
126. Cifelli C, Rose RA, Zhang H, et al. RGS4 regulates parasympathetic signaling and heart rate control in the sinoatrial node. *Circ Res*. 2008;103(5):527–535.
127. Wydeven N, Posokhova E, Xia Z, Martemyanov KA, Wickman K. RGS6, but not RGS4, is the dominant regulator of G protein signaling (RGS) modulator of the parasympathetic regulation of mouse heart rate. *J Biol Chem*. 2014;289(4):2440–2449.
128. Yang J, Huang J, Maity B, et al. RGS6, a modulator of parasympathetic activation in heart. *Circ Res*. 2010;107(11):1345–1349.
129. Posokhova E, Wydeven N, Allen KL, Wickman K, Martemyanov KA. RGS6/Gbeta5 complex accelerates IKACh gating kinetics in atrial myocytes and modulates parasympathetic regulation of heart rate. *Circ Res*. 2010;107(11):1350–1354.

130. Raveh A, Schultz PJ, Aschermann L, et al. Identification of PKC activation as a novel mechanism for RGS2 protein up-regulation through phenotypic screening of natural product extracts. *Mol Pharmacol.* 2014;86(4):406–416.
131. Xie Y, Wolff DW, Wei T, et al. Breast cancer migration and invasion depend on proteasome degradation of regulator of G-protein signaling 4. *Cancer Res.* 2009;69(14):5743–5751.
132. Rogers JH, Tamirisa P, Kovacs A, et al. RGS4 causes increased mortality and reduced cardiac hypertrophy in response to pressure overload. *J Clin Invest.* 1999;104(5):567–576.
133. Jones TL. Role of palmitoylation in RGS protein function. *Methods Enzymol.* 2004;389:33–55.
134. Burchett SA. In through the out door: nuclear localization of the regulators of G protein signaling. *J Neurochem.* 2003;87(3):551–559.
135. Abramow-Newerly M, Roy AA, Nunn C, Chidiac P. RGS proteins have a signalling complex: interactions between RGS proteins and GPCRs, effectors, and auxiliary proteins. *Cell Signal.* 2006;18(5):579–591.
136. Chen B, Siderovski DP, Neubig RR, Lawson MA, Trejo J. Regulation of protease-activated receptor 1 signaling by the adaptor protein complex 2 and R4 subfamily of regulator of G protein signaling proteins. *J Biol Chem.* 2014;289(3):1580–1591.
137. Blazer LL, Roman DL, Chung A, et al. Reversible, allosteric small-molecule inhibitors of regulator of G protein signaling proteins. *Mol Pharmacol.* 2010;78(3):524–533.
138. Cheng WL, Wang PX, Wang T, et al. Regulator of G-protein signalling 5 protects against atherosclerosis in apolipoprotein E-deficient mice. *Br J Pharmacol.* 2014 Nov 3. http://dx.doi.org/10.1111/bph.12991 [Epub ahead of print].

CHAPTER SEVEN

Two for the Price of One: G Protein-Dependent and -Independent Functions of RGS6 *In Vivo*

Adele Stewart[1], Biswanath Maity, Rory A. Fisher[2]
Department of Pharmacology, University of Iowa Carver College of Medicine, Iowa City, Iowa, USA
[2]Corresponding author: e-mail address: rory-fisher@uiowa.edu

Contents

Abstract

Regulator of G protein signaling 6 (RGS6) is unique among the members of the RGS protein family as it remains the only protein with the demonstrated capacity to control G protein-dependent and -independent signaling cascades *in vivo*. RGS6 inhibits signaling mediated by γ-aminobutyric acid B receptors, serotonin 1A receptors, μ opioid receptors, and muscarinic acetylcholine 2 receptors. RGS6 deletion triggers distinct behavioral phenotypes resulting from potentiated signaling by these G protein-coupled receptors namely ataxia, a reduction in anxiety and depression, enhanced analgesia, and increased parasympathetic tone, respectively. In addition, RGS6 possesses potent proapoptotic and growth suppressive actions. In heart, RGS6-dependent reactive

[1] Present address: Department of Pharmacology, Vanderbilt University, Nashville, Tennessee, USA.

Progress in Molecular Biology and Translational Science, Volume 133
ISSN 1877-1173
http://dx.doi.org/10.1016/bs.pmbts.2015.03.001

oxygen species (ROS) production promotes doxorubicin (Dox)-induced cardiomyopathy, while in cancer cells RGS6/ROS signaling is necessary for activation of the ataxia telangiectasia mutated/p53/apoptosis pathway required for the chemotherapeutic efficacy of Dox. Further, by facilitating Tip60 (*trans*-acting regulator protein of HIV type 1-interacting protein 60 kDa)-dependent DNA methyltransferase 1 degradation, RGS6 suppresses cellular transformation in response to oncogenic Ras. The culmination of these G protein-independent actions results in potent tumor suppressor actions of RGS6 in the murine mammary epithelium. This work summarizes evidence from human genetic studies and model animals implicating RGS6 in normal physiology, disease, and the pharmacological actions of multiple drugs. Though efforts by multiple laboratories have contributed to the ever-growing RGS6 oeuvre, the pleiotropic nature of this gene will likely lead to additional work detailing the importance of RGS6 in neuropsychiatric disorders, cardiovascular disease, and cancer.

1. INTRODUCTION

1.1 R7 Family RGS Proteins

The R7 subfamily of RGS proteins, which includes RGS6, RGS7, RGS9, and RGS11, shares a characteristic three-domain structure. The RGS domain confers functional GTPase-activating protein (GAP) activity directed specifically toward $G\alpha_{i/o}$.[1] The N-terminal disheveled, EGL-10, pleckstrin homology (DEP)/DEP helical extension (DHEX) domain is best known for mediating interaction between R7 family RGS proteins and two accessory proteins: R7 family-binding protein (R7BP) and RGS9 anchor protein (R9AP).[2] R7BP is widely expressed in the nervous system with detectable expression in the mammalian brain, spinal cord, and retina[3] and, unlike R9AP, which associates exclusively with RGS9-1 and RGS11, is capable of binding to all R7 family members.[4] R7BP is reversibly palmitoylated and, as a result, mediates shuttling of R7 family RGS proteins between the nucleus and plasma membrane.[5] In neurons, R7BP is required for recruitment of RGS9-2 to postsynaptic densities, specialized electron-dense regions on neuronal cell membranes that contain large quantities of neurotransmitter receptors and signal transduction machinery.[6] RGS9-2 is also stabilized through its interaction with R7BP in the striatum, though R7BP expression does not appear to be required for the stability of RGS6, RGS7, or RGS11.[7,8] Mice lacking R7BP exhibit ataxia and potentiation of opioid analgesia (Table 1), phenotypes attributed to loss of membrane targeting of one or more R7 family RGS proteins.[18–20]

The G gamma subunit-like (GGL) domain, located between the N-terminal DEP/DHEX domains and the C-terminal RGS domain,

Table 1 Reported Phenotypes of R7BP$^{-/-}$ and Gβ5$^{-/-}$ Mice

Mouse Model	Phenotype	Year	Reference(s)
$G\beta_5^{-/-}$	Stunted growth	2003	9
	Prolonged photoresponses and defective light adaptation	2003	10
	Altered retinal morphology	2007	11
	Potentiation of M2R-induced GIRK current	2010	12
	Increased sensitivity to the sedative effects of $GABA_BR$ activation	2010	13
	Reduced body weight and resistance to diet-induced obesity, hyperactivity	2011	14
	Impaired brain development, ataxia, and motor coordination defects	2011	15
	Deficits in motor learning and sensitization to new environments, hyperactivity	2012	16
	Enhanced striatal adenylyl cyclase V activity	2012	17
$G\beta_5^{+/-}$	Increased body weight and sensitivity to diet-induced obesity, hyperactivity	2011	14
R7BP$^{-/-}$	Ataxia, increased sensitivity to locomotor-stimulatory effects of morphine	2010	18,19
	Enhanced opiate analgesia and tolerance	2012	19,20
	Potentiation of $GABA_BR$–GIRK current	2012	20

Note: Studies are presented in chronological order with associated references.

facilitates interaction between R7 family members and the atypical Gβ subunit $G\beta_5$.[21–23] Complex formation between R7 family members and $G\beta_5$ is required for the stable expression of both proteins as genetic ablation of the *GNB5* gene results in loss of functional expression of RGS6, RGS7, RGS9, and RGS11.[9] Stabilization of R7 RGS proteins also requires interaction between $G\beta_5$ and the DEP domain on an interface highly conserved between mammalian R7 family RGS proteins and the ancestral R7 family-like RGS protein EGL-10 expressed in *C. elegans*. Mutation of these residues leads to proteasomal degradation of both components of the R7 RGS protein/$G\beta_5$ complex.[24] $G\beta_5^{-/-}$ mice have aided in identification of physiological processes modulated by R7 family RGS proteins *in vivo* including phototransduction, motor coordination, locomotor activity, cardiac automaticity, metabolism, and development (Table 1). However, studies in $G\beta_5^{-/-}$ mice have failed to identify which R7 family member(s) mediate each phenotype.

1.2 The RGS6 Gene, mRNA, and Protein

The Fisher laboratory first identified and cloned RGS6, the founding member of the R7 family, and later described the existence of complex alternative splicing of regulator of G protein signaling 6 (RGS6) in brain.[25] Splicing of two primary RGS6 transcripts yields 36 distinct isoforms, each with a distinct protein structure (Fig. 1). This variation influences the ability of the individual isoforms to associate with $G\beta_5$ and R7BP and, as a result, impacts their stability and subcellular localization. More specifically, RGS6S isoforms lack the DEP/DHEX domains and, thus, are not subject to R7BP-mediated plasma membrane targeting. Similarly, RGS6 isoforms lacking exon 13 (GGL domain) are unable to bind $G\beta_5$ resulting in protein destabilization. The largest species are the 18 variants of the RGS6L (long) form. All RGS6L isoforms possess the hallmark RGS and DEP/DHEX domains, but differ in having one of nine distinct C-terminal sequences and by the presence or absence of a complete GGL domain (Fig. 1).

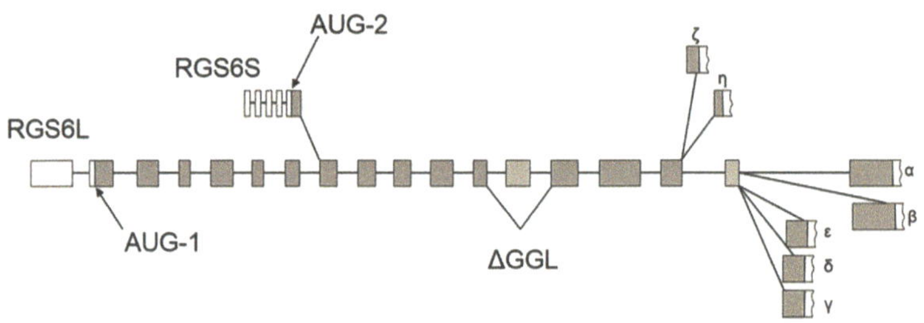

Figure 1 Schematic depicting the splicing events occurring in the human *RGS6* gene to generate the 36 known RGS6 isoforms. Exons are shown as filled boxes and noncoding sequences as empty boxes. Two primary transcripts with alternate translation start sites (AUG-1 and AUG-2) generate RGS6L or RGS6S forms, the latter lacking a large portion of the N-terminus including the disheveled, EGL-10, pleckstrin homology/DEP helical extension (DEP/DHEX) domains. Retention or skipping of exon 13 generates transcripts with or without a GGL domain (ΔGGL), respectively. Extensive 3′ splicing generates seven different 3′ ends. Use of alternate splice sites in exon 20 generates α and β species while γ, ε, and δ arise from the use of three alternate sites in exon 19. Splicing of exon 16 or 17 to an alternate splice site in exon 18 generates η and ζ, respectively. Additionally, α and β isoforms may contain (1) or lack (2) exon 17. *Image adapted from Ref. 25.* (See the color plate.)

RGS6 mRNA expression was first detected in mammalian brain[26] with transcript levels also evident in human heart.[27] Using an antibody raised against the unique N-terminus of RGS6, nervous system expression of multiple RGS6 immunoreactive bands was confirmed in multiple brain regions including the cortex, hippocampus, cerebellum, striatum, and the ventral tegmental area (Fig. 2A). RGS6 expression was also found in multiple peripheral tissues including heart and lung (Fig. 2B) as well as bladder, omental fat, and the gastrointestinal tract (data not shown). Importantly, RGS6 expression is lost in RGS6$^{-/-}$ mice (Fig. 2).

Based on their canonical actions in G protein regulation, it is assumed that localization of RGS proteins to the plasma membrane is essential to their function. However, studies of RGS6 in heterologous expression systems indicate that RGS6 splice forms exhibit differences in their subcellular distribution and can even traffic to the nucleus. More specifically, RGS6S isoforms localize to the nucleus, while RGS6L isoforms remain in the cytoplasm.[25] In this cell culture system, Gβ_5 coexpression promoted nuclear localization of RGS6L protein species. RGS6 and RGS7 are detectable in both membrane and cytosolic fractions in neural lysates with RGS7 preferably found in the membrane fraction and RGS6 in the cytosol (Fig. 2C). Though experiments to establish whether RGS6 exists in the nucleus have yet to be performed in native tissue, RGS6 does interact with nuclear

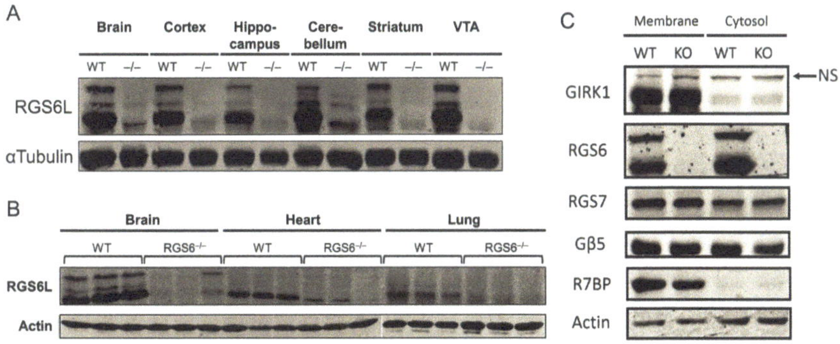

Figure 2 RGS6L protein expression. (A) Immunoblotting for RGS6L using an antibody developed in our laboratory against the N-terminus of RGS6 was performed in WT and RGS6$^{-/-}$ mice for each tissue listed. Multiple immunoreactive bands were detected that were completely gone in the RGS6$^{-/-}$ mice (a nonspecific band below RGS6 is present in some tissues). αTubulin was used as a loading control. (B) RGS6 immunoreactivity in the brain, heart, and lung of WT and RGS6$^{-/-}$ mice ($n=3$). Actin was used as a loading control. (C) Immunoreactivity of GIRK1, RGS6, RGS7, Gβ_5, R7BP, and actin in cytosolic and membrane fractions from whole brain tissue lysates from WT and RGS6$^{-/-}$ (KO) mice.

proteins in mouse brain providing indirect evidence that RGS6 is capable of traveling into the nucleus *in vivo*.[28]

1.3 *In Vitro* Functions of RGS6

Prior to the acquisition and development of $RGS6^{-/-}$ mice, the Fisher laboratory ascribed a few functions to RGS6 *in vitro*. First, thermal, proteotoxic or transcription-linked stress is sufficient to induce trafficking of RGS6 to the nucleoli, where RGS6 would be incapable of functioning as a GAP for Gα subunits.[29] Interestingly, the RGS domain of RGS6 was primarily responsible for promoting stress-induced nucleolar accumulation, one of the first identified functions for an RGS domain outside of G protein regulation. While the functional consequence of this trafficking event in mammalian physiology remains unclear, there is some evidence for alterations of RGS6 expression in disease where cellular stress pathways may be activated such as ischemic stroke.[30]

In neuron-like cells, RGS6 interacts with the neuronal growth-associated protein stathmin-like 2 (STMN2, SCG10) and promotes neurite outgrowth and cell differentiation.[31] Specifically, RGS6 coexpression potentiated the ability of SCG10 to disrupt the microtubule network. SCG10 binds to the GGL domain of RGS6, though it remains unclear whether this interaction competes with $G\beta_5$ binding. Further, an RGS6 mutant incapable of interacting with G proteins ($RGS6^{N401V}$) was also able to influence cell differentiation identifying an entirely novel, G protein-independent action of RGS6. Recent evidence demonstrating that RGS6 is upregulated by several fold following differentiation of human neuronal progenitors underscores the potential importance of RGS6 in this process.[32] Little is known regarding additional functions of SCG10 *in vivo* outside of a known role in neuronal differentiation and neurite outgrowth in cultured cells. However, it has been implicated in both nerve degeneration and regeneration following injury[33,34] and neuronal cell migration during development.[35] The impact of RGS6 on these SCG10-dependent processes has not been investigated.

RGS6 also binds to DNA methyltransferase 1 (Dnmt1) via an interaction with DNA methyltransferase 1-associated protein 1 (DMAP1) leading to suppression of the transcriptional repressor actions of DMAP1.[36] DMAP1 also binds to the GGL domain of RGS6 on an interface not required for the $G\beta_5$–RGS6 association.[36] The importance of the RGS6–Dnmt1 interaction is discussed in greater detail later in this chapter.

2. REGULATION OF G PROTEIN SIGNALING BY RGS6 *IN VIVO*

2.1 G Protein-Dependent RGS6 Functions in Brain

There is now substantial evidence that RGS6 is expressed in multiple neuronal circuits including cortex, hippocampus, cerebellum, and the mesolimbic system (Fig. 2A) where it functions as a critical negative regulator of signaling by multiple G protein-coupled receptors (GPCRs).

2.1.1 μ Opioid Receptor-Mediated Analgesia

Initial work seeking to identify GPCRs regulated by R7 family members *in vivo* utilized intracerebroventricular injection of antisense oligonucleotides targeting RGS6, RGS7, RGS9-2, RGS11, or Gβ_5 to study the role of the individual RGS proteins in μ opioid receptor (MOR) tachyphylaxis and acute tolerance.[37,38] RGS6 knockdown enhanced the acute analgesic efficacy of morphine, but failed to impact drug tolerance. Unlike results seen with RGS7-targeted oligonucleotides, RGS6 appeared to impact MOR- and not δ opioid receptor-dependent analgesia.[37] Opioids also appear to up- or downregulate R7 family RGS proteins differentially depending on the time following treatment and the specific brain region evaluated.[39] For example, RGS6 expression decreased following chronic morphine treatment in the periaqueductal gray matter but not in the cortex, thalamus, or striatum. Clearly, RGS6 normally functions to terminate MOR signaling. However, these studies achieved only partial RGS6 depletion, and thus, the extent of RGS6 action was likely unrepresented. The development of complete RGS6$^{-/-}$ mice would facilitate investigations into the net importance of RGS6-mediated GPCR inhibition in multiple physiological processes.

2.1.2 GABA$_B$R-Dependent Motor Coordination

Though mice lacking RGS6 exhibit no gross growth retardation, reproductive deficiencies or obvious behavioral abnormalities, when challenged with GPCR-targeted drugs or in specific paradigms differences begin to arise. The first study of RGS6-modulated behaviors investigated the impact of *RGS6* gene deletion on motor movement.[40] RGS6$^{-/-}$ mice exhibit gait and stance abnormalities and impaired performance on the rotarod, the latter indicative of ataxia due to cerebellar dysfunction. Mice lacking RGS6 are also sensitized to the ataxic effects of baclofen, a selective agonist for

$GABA_BRs$. Further, inhibition of $GABA_BR$ signaling through administration of a selective $GABA_BR$ antagonist rescued the ataxic phenotype of $RGS6^{-/-}$ mice without impacting the performance of wild-type animals. These results provided the first evidence that RGS6 suppresses $GABA_BR$ signaling activated by endogenous GABA in the brain. Unsurprisingly, RGS6 is highly expressed in the cerebellum, particularly in the granule cell layer. In isolated cerebellar granule neurons, RGS6 functions to inactivate $G\alpha_{i/o}$ released from $GABA_BRs$ and induce resequestration of $G\beta\gamma$ and termination of downstream activation of G protein-activated inwardly rectifying potassium (GIRK) channels (Fig. 3). Loss of rapid GIRK channel deactivation could prolong GABA-mediated membrane depolarization in these neurons resulting in an imbalance of excitatory and inhibitory neurotransmission in the cerebellum, an aberration known to manifest in gross cerebellar dysfunction. It is important to note that granule cell inhibition in the adult cerebellum is dominated by tonic $GABA_A$-mediated responses without a reported $GABA_B$-dependent component.[41] Thus, it remains to be determined whether the ataxic phenotype of $RGS6^{-/-}$ mice is uniquely determined by loss of $GABA_BR$-mediated GIRK current deactivation.

In cerebellum, RGS6 forms a complex with known binding partners $G\beta_5$ and R7BP. Based on its known role in facilitating membrane localization of R7 family RGS proteins,[42] R7BP binding likely recruits RGS6 to

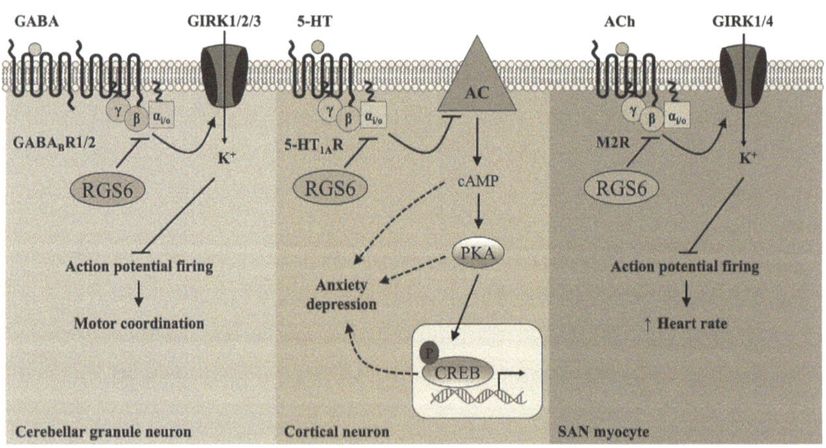

Figure 3 Schematic outlining the role of RGS6 in GPCR signaling. RGS6 functions to limit the extent of $GABA_BR$-mediated GIRK channel activation in cerebellar granule cells; $5\text{-}HT_{1A}R$-mediated inhibition of AC in cortical neurons and M2R-mediated GIRK current in SAN myocytes. In this way, RGS6 controls motor behavior, baseline anxiety and depression, and heart rate *in vivo*. (See the color plate.)

the cell surface where RGS6–Gβ_5 complexes could mediate rapid GIRK channel inactivation.[43] Though complex formation between R7 family RGS proteins, Gβ_5, and GIRK1 has been independently confirmed by multiple laboratories,[13,20,44,43] there remains only a single report of an interaction in native tissues.[13] Efforts to confirm a direct interaction between RGS6 and GIRK1 or GIRK2 in brain were unsuccessful though RGS6 along with GIRK1, GIRK2, and GABA$_B$R2 is uniformly localized throughout the soma and neurites in cerebellar granule neurons, suggesting that RGS6 is positioned within neurons to modulate the GABA$_B$R–GIRK channel signaling axis despite the lack of its direct binding to GIRK channels.[40] Interestingly, RGS6$^{-/-}$ hippocampal neurons unlike cerebellar granule neurons exhibit normal baclofen-induced GIRK channel kinetics.[45] Instead, RGS7 appears to be the primary regulator of the GABA$_B$R–GIRK axis in this cell population.[44,45] Thus, it is entirely possible that R7 family RGS proteins modulate GABA$_B$R–GIRK signaling through distinct temporal, spatial, and biochemical mechanisms.

Indeed, maintenance of Gβ_5 expression in RGS6$^{-/-}$ animals suggests that other R7 family members are also expressed in cerebellum and, thus, able to stabilize Gβ_5 in the absence of RGS6. The fact that loss of RGS6 alone results in deficits in motor coordination and movement suggests that other R7 RGS family members, though expressed in cerebellum, are unable to compensate for loss of RGS6. In fact, R7 family members have been implicated both in regulation of GABA$_B$R signaling as well as motor coordination. R7BP$^{-/-}$ mice also exhibit impaired performance on the rotarod, though this phenotype is primarily due to loss of striatal-specific expression of RGS9-2.[18] Similarly, mice lacking Gβ_5 exhibit ataxia; however, this phenotype likely results from aberrant cerebellar development, a phenomenon that does not occur in RGS6$^{-/-}$ mice.[15] Investigation into the motor phenotypes of other R7 family-specific knockouts might provide additional insight into how this group of G protein regulators contributes to cerebellar function.

2.1.3 5-HT$_{1A}$R Activation in Anxiety and Depression

Targeting 5-HT bioavailability with selective 5-HT reuptake inhibitors (SSRIs) remains the most widely used treatment for mood disorders. However, their limited efficacy, delayed onset of action, and side effects restrict their clinical utility. Endogenous RGS proteins have been implicated as key inhibitors of serotonin (5-HT) 1A receptors (5-HT$_{1A}$Rs), whose activation is believed to underlie the beneficial effects of SSRIs.[46] RGS6 has now been identified as the critical negative regulator of 5-HT_{1A}R-dependent antidepressant actions. RGS6 is enriched in hippocampal and cortical neurons, 5-HT_{1A}R-expressing cells implicated in mood disorders,[47,48] and mice lacking RGS6 exhibit spontaneous anxiolytic and antidepressant behavior rapidly and completely reversible by 5-HT_{1A}R blockade.[49] While the phenotype of RGS6 heterozygotes is indistinguishable from WT mice, partial RGS6 deficiency elicits a remarkable sensitization to the antidepressant actions of the SSRI fluvoxamine and 5-HT_{1A}R agonist 8-OH-DPAT. These data suggest that partial loss or inhibition of RGS6 is sufficient to potentiate the actions of serotonergic drugs.

5-HT_{1A} heteroreceptors activate diverse signaling in postsynaptic neurons located in both hippocampus and cortex and are believed to mediate the antidepressant and antianxiety actions of serotonin.[47,50] Loss of RGS6 potentiates 5-HT_{1A}R–$G\alpha_i$-dependent adenylyl cyclase (AC) inhibition *in vivo* and in isolated cortical neurons resulting in a reduction in protein kinase A (PKA) substrate phosphorylation and phosphorylation of the transcription factor cyclic AMP response element binding protein (CREB). Indeed, RGS6 loss appears to selectively enhance heteroreceptor and not autoreceptor populations as RGS6 heterozygotes are sensitized to the antidepressant actions of 5-HT_{1A}R stimulation, but 8-OH-DPAT-induced hypothermia, an effect dependent on autoreceptor action,[51] occurs with equal potency in $RGS6^{-/-}$, $RGS6^{+/-}$, and WT mice.[49] Second, only 5-HT_{1A} heteroreceptors couple to $G\alpha_i$ and induce inhibition of AC[52] and the antidepressant phenotype of $RGS6^{-/-}$ mice is dependent on AC inhibition as it is completely reversed by forskolin.[49] In short, RGS6 functions as the gatekeeper for 5-HT-mediated anxiolytic and antidepressant actions due primarily to its ability to completely block 5-HT_{1A}R-mediated inhibition of AC (Fig. 3).

Mice expressing an RGS-insensitive mutation in $G\alpha_{i2}$ also exhibit an antidepressant and anxiolytic phenotype correlated with alterations in glycogen synthase kinase 3β (GSK3β) phosphorylation in hippocampus and cortex.[46] Given studies identifying GSK3β as an important molecular target of 5-HT,[53] it was postulated that increased GSK3β phosphorylation contributed to the anxiolytic phenotype of $G\alpha_{i2}$(G148S) mutants. $RGS6^{-/-}$ mice show an identical phenotype in the absence of increased phospho-GSK3β.[49] This result implies both that increased GSK3β phosphorylation is dispensable for the ability of 5-HT_{1A}Rs to reduce anxiety and depression and that another RGS protein is responsible for modulation of 5-HT_{1A}R-mediated

GSK3β inactivation. RGS6 appears to selectively regulate a specific 5-HT_{1A}R-effector pathway, namely the $G\alpha_i$-AC–PKA–CREB axis, without impacting other targets (e.g., GSK3β). Indeed, the ability of forskolin to reverse the antidepressant phenotype of $RGS6^{-/-}$ mice suggests that Gβγ effectors (GSK3β, MAPK, Akt) are not required for the ability of 5-HT_{1A}Rs to reduce depression.[49]

The selective modulation of $G\alpha_i$- and not Gβγ-dependent signaling by RGS6 in hippocampus and cortex is inconsistent with RGS6 solely functioning as a GAP, which should impact the signaling of both components of the heterotrimeric G protein complex. There are a few possible explanations for this conundrum. First, it is possible that RGS6 influences the signaling of a distinct $G\alpha_i$ isoform. If true, this could explain why loss of RGS protein-mediated regulation of $G\alpha_{i2}$ effects GSK3β phosphorylation while loss of RGS6 impacts the AC/cAMP/PKA/CREB axis. Such a result would imply that $G\alpha_i$ isoforms exhibit selectivity in effector activation, a phenomenon that has been observed for $G\alpha_{i/o}$ isoform–GPCR coupling.[54] This could result from differences in the abundance or Gβγ composition of $G\alpha_{i1}$ vs. $G\alpha_{i2}$ containing heterotrimeric complexes as prior studies have shown that the composition of Gβγ subunits can be a critical determinant in their effector coupling.[55–57] Second, RGS6 may specifically interact with macromolecular complexes containing the 5-HT_{1A}R–$G\alpha_i$-AC signaling components. Though such a complex has not been demonstrated for the 5-HT_{1A}R specifically, emerging evidence suggests that GPCRs, G proteins, and their effectors localize to specialized microdomains within neurons believed to facilitate fast receptor–effector coupling as well as signaling specificity.[58] Interestingly, the 5-HT_{1A}R does bind to the GIRK3 channel subunit *in vitro*, an interaction dependent on RGS protein expression.[59] Also, recent evidence suggests that R7 family RGS protein RGS9-2 binds to AC V in striatum,[17] implying that RGS9-2 and perhaps the other R7 family members may interact with AC isoforms expressed elsewhere in the nervous system. Indeed, there are numerous examples of complex formation between RGS proteins, GPCRs, G proteins, and various effectors lending veracity to this second postulate.[60,61]

R7 RGS protein family members including RGS7 are also expressed in hippocampus and cortex. However, the members of this small protein subfamily appear to be functionally divergent. For example, RGS6 appears to regulate 5-HT_{1A}Rs in the hippocampus, which may contribute to anxiety- and depression-related behaviors and the actions of serotonergic drugs, while RGS7 appears to impact $GABA_BR$ signaling involved in hippocampal synaptic plasticity and contextual memory.[45] No such deficits were observed in mice lacking RGS6.[45] Though RGS6 is capable of regulating $GABA_BRs$ in cerebellar neurons, RGS7 appears to be the dominant inhibitor of these receptors in hippocampal neurons. Similarly, RGS7 is unable to regulate $5\text{-}HT_{1A}R$-mediated inhibition of AC *in vitro*[62] consistent with the primary function of RGS6 in regulating this receptor population. Mice lacking RGS9 exhibit no changes in anxiety, though their behavior has only been evaluated in the open field test.[63] Because mice lacking RGS7 or RGS11 have not been evaluated in the behavioral paradigms described in this work, it is unknown whether these R7 family members have any role in modulating anxiety and depression.

2.2 G Protein-Dependent RGS6 Functions in Heart

Acetylcholine (ACh) released from the vagus nerve binds to muscarinic acetylcholine 2 receptors (M2Rs) triggering the activation of GIRK channels, which represent the primary determinants of ACh-activated potassium current (I_{KACh}), which promotes membrane hyperpolarization and inhibition of action potential firing in the pacemaker and electrically conducting portions of the heart.[64] For many years, researchers observed a discrepancy between the gating kinetics of GIRK channel current in atrial myocytes compared to that of GPCRs and GIRK subunits reconstituted in heterologous expression systems, wherein channel deactivation is up to 40-fold slower.[65,66] Remarkably, expression of a single RGS protein (RGS4) was able to restore GIRK channel activation and deactivation kinetics to those observed in intact myocytes.[67] Though numerous RGS proteins are expressed in heart,[68] studies in individual knockout mice have confirmed that RGS6 is the primary regulator parasympathetic stimulation of heart.[12,69,70] RGS6 expression is enriched in the sinoatrial and atrioventricular nodes, the primary pacemaking portions of the myocardium. Loss of RGS6 leads to enhanced M2R-dependent GIRK channel activation, inhibition of sinoatrial node (SAN) action potential firing, and bradycardia.[12,69]

Tightly regulated control of SAN pacemaker cells by inputs from the autonomic nervous system maintains rhythmic cardiac contractions and a regular heartbeat. Fluctuations in parasympathetic nerve activity under resting conditions are the primary determinant of heart rate variability and such measures provide an additional index of the magnitude of parasympathetic modulation of heart rate.[71] Consistent with its demonstrated role in regulating I_{KACh} generated following vagal stimulation of the heart, *RGS6* ablation results in irregular heart rate and increased susceptibility to atrial fibrillation. In addition, this same study identified a novel variant in the *RGS6* gene correlated with increased heart rate variability.[70] Reactivation of vagal efferents is required for the rapid, exponential decline in heart rate following exercise. Thus, the identification of an additional single nucleotide polymorphism (SNP) in the *RGS6* gene associated with deficits in HR recovery after exercise[72] further underscores the importance of RGS6 in controlling parasympathetic stimulation of heart, especially given faster post-exercise heart rate recovery is linked to lower risk of morbidity and mortality in chronic heart failure.[73,74] Taken together, these studies indicate that RGS6 deficiency may be a double-edged sword, possibly improving health outcomes in heart failure but increasing risk for autonomic dysfunction-related arrhythmias.

The potential contribution of other RGS proteins in regulation of M2R signaling cascade in heart is controversial with conflicting reports regarding the involvement of RGS4.[68,75,76] Though Cifelli *et al.* reported in 2008 that RGS4 ablation resulted in increased bradycardic responses to parasympathetic agonists,[76] a recent study reported conflicting results demonstrating no change in parasympathetic activity in mice lacking RGS4.[75] Similarly, the increased sensitivity to M2R-mediated GIRK channel activation and inhibition of action potential firing in sinoatrial myocytes isolated from RGS4$^{-/-}$ mice was not observed.[75] The discrepancy between these findings is odd, given these independent labs utilized the same RGS4$^{-/-}$ mouse line. Of note, RGS4 mRNA is enriched in the heart, but there is little to no detectable RGS4 protein expression.[69,77] The lack of correlation between RGS4 mRNA and protein levels observed[78] may be due to a destabilizing N-terminal cysteine residue whose arginylation targets RGS4 for degradation through the ubiquitin-dependent N-end rule pathway.[79,80] Because RGS6 expression is readily detectable in pacemaking cells of the heart, it makes sense that loss of this protein would have a more dramatic effect. RGS4 clearly plays a role, however, as RGS4 deficiency was able to partially rescue the deficits in M2R–GIRK channel kinetics and cholinergic agonist-induced bradycardia observed in RGS6$^{-/-}$ mice.[75] Nevertheless, RGS6 appears to be the predominant brake on vagal stimulation of the heart (Fig. 3).

3. G PROTEIN-INDEPENDENT RGS6 SIGNALING ACTIONS

RGS6 is unique among R7 family members, because, though capable of modulating multiple GPCRs (Fig. 3), RGS6 also participates in multiple G protein-independent signaling cascades.

3.1 Apoptosis and Growth Suppression

The first evidence linking RGS6 to carcinogenesis was published in 2004 demonstrating a strong association between an SNP in the *RGS6* gene with a reduction in bladder cancer risk especially in smokers. This polymorphism mapped to the 3′ untranslated region of RGS6 mRNA and resulted in a significant increase in translation.[81] This led to the novel hypothesis that RGS6 possessed some growth suppressive or cytotoxic actions in cancer cells. In keeping with this observation, RGS6 expression, normally localized to the breast ductal epithelium, is reduced in breast tumor samples isolated from human cancer patients with loss of RGS6 protein correlating with increasing tumor malignancy.[82,83] Interestingly, RGS6 expression is lost independent of tumor molecular subclassification (e.g., estrogen receptor, ER; progesterone receptor, PR; or human epidermal growth factor receptor 2, HER2 immunoreactivity) indicating a more universal role for RGS6 in suppressing breast tumor initiation, progression, and/or metastasis.

Indeed, exogenous expression of RGS6L in breast cancer cell lines, which express little to no endogenous RGS6, dramatically compromised cell survival, triggered cell cycle arrest in the G1/S phase, and impeded colony formation.[82] Two potential mechanisms were proposed to explain these phenomena. First, RGS6 altered the activity or expression of cyclins D1 and E and retinoblastoma protein, critical regulators of cell cycle progression. Second, RGS6 activated the intrinsic apoptosis pathway characterized by alterations in the Bax/Bcl-2 ratio, loss of mitochondrial membrane potential, cytochrome *c* release into the cytosol, and cleavage of initiator caspase-9 and executioner caspase-3. The ability of RGS6 to promote apoptosis required the generation of reactive oxygen species (ROS), specifically superoxide and hydrogen peroxide. Though GPCR signaling has been heavily implicated in carcinogenesis,[84] the ability of RGS6 to promote apoptosis and suppress growth in cancer cells did not require its GAP activity indicating an entirely novel function for an RGS protein. In fact, this was the first study describing a role for an RGS protein in cell death pathways (Fig. 4). Though this work generated compelling evidence demonstrating

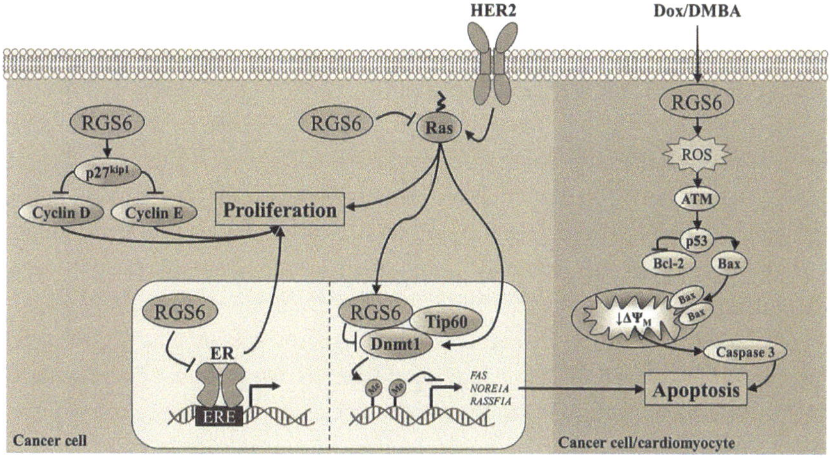

Figure 4 Schematic outlining the role of RGS6 in G protein-independent signaling. In cancer cells, RGS6 is a multifunctional tumor suppressor capable of simultaneously promoting cell cycle arrest and blocking progrowth signals downstream of HER2 and ER. Further, the ability of RGS6 to promote Tip60-mediated Dnmt1 acetylation and degradation is essential to combating Ras-induced oncogenesis. In addition, RGS6 is a critical upstream modulator of ROS-dependent activation of the ATM/p53/apoptosis pathway in response to doxorubicin in cancer cells and heart. (See the color plate.)

that RGS6 is a potential new tumor suppressor, evidence from an *in vivo* system was required to confirm this postulate.

3.2 Doxorubicin-Induced Cytotoxicity

3.2.1 Cancer Cells

The anthracycline doxorubicin (Dox) is among the most widely used chemotherapeutic agents. Dox-induced double-stranded DNA breaks result from either inhibition of DNA topoisomerase II or the generation of ROS and trigger the DNA damage response (DDR) leading to apoptotic cell death.[85–87] The ataxia telangiectasia-mutated (ATM) protein kinase plays a critical role in sensing DNA damage and in orchestrating the signaling output of the DDR including induction of the proapoptotic tumor suppressor p53.[87] Dox induces dramatic upregulation of RGS6 in cancer cells and mouse embryonic fibroblasts (MEFs).[88] The ability of Dox to activate the ATM/p53/apoptosis pathway and suppress cell growth is also almost completely absent in MEFs isolated from RGS6$^{-/-}$ mice or cancer cells where RGS6 expression was reduced using interfering RNA.[82,88] The observation that expression of RGS6L but not the GAP-defective mutant RGS6L^{N401V} was able to restore p53 phosphorylation and upregulation in RGS6 null cells treated with Dox indicates that RGS6 promotes p53 activation via a G protein-independent mechanism.[88] Additional experiments indicated that RGS6 functions upstream of ATM and is able to promote ATM activation by the recently described, ROS-dependent oxidative mechanism rather than by DNA damage.[89]

These experiments expanded on previous work in heterologous expression systems and underscored the potential importance of RGS6-mediated ROS generation in the actions of one of the most effective chemotherapeutic agents (Fig. 4). Of note, while RGS6 functions upstream of p53 in Dox-induced activation of the DDR, p53 is not specifically required for the ability of RGS6 to promote apoptosis as exogenous expression of RGS6 leads to apoptosis with equivalent efficacy in cells harboring mutant p53.[82] Thus, RGS6 may be capable of activating multiple intracellular signaling cascades that converge on cell death pathways. The common thread in these distinct actions is the ability of RGS6 to induce oxidative stress. Though the mechanism whereby RGS6 promotes ROS accumulation remains unclear, these studies indicate that the G protein-independent actions of RGS6 might represent a critical barrier to cancer cell propagation. This assertion is consistent with the gradual downregulation of RGS6 observed in human breast cancer patient samples with increasing grade,[82,83] which would be expected to increase cell proliferation and impair the cytotoxic efficacy of Dox. The participation of RGS6 in proapoptotic signaling induced by additional chemotherapeutic drugs remains unknown.

3.2.2 Cardiomyopathy

The major limitation for the clinical use of Dox is its life-threatening cardiac actions that include cardiomyopathy and heart failure.[90] Patients receiving cumulative doses of Dox of 700 mg/m^2 have a 48% risk of developing heart failure.[91] More than 50% of childhood cancer survivors have been treated with anthracyclines and the mortality rate from cardiac death for 30-year survivors is 15 times higher than expected.[92,93] Apoptotic death of cardiomyocytes, considered irreversible due to their terminal differentiation, is a prominent pathological response to heart injury and stress and underlies the morphological and functional changes in the failing myocardium. Both p53 and oxidative stress have been implicated in cardiac cell death following Dox-mediated heart injury.[94,95] In keeping with critical role of RGS6 in Dox-induced cell death and the robust expression of RGS6 in the heart, it is unsurprising that $RGS6^{-/-}$ mice are largely protected from the cardiac pathogenic response to Dox including activation of the

ATM/p53 signaling cascade, apoptotic cell death, cardiac microfilament disarray, and loss of ventricular function.[88,96] Dox also increases RGS6 expression in isolated ventricular myocytes where RGS6-dependent ROS generation is required for Dox-mediated apoptosis.[96] Together these data support a model in which RGS6 upregulation by Dox in heart induces ROS generation, which in turn promotes ATM/p53 apoptosis signaling via oxidative activation of ATM (Fig. 4).

3.3 RGS6 in Carcinogenesis

3.3.1 Ras-Induced Cellular Transformation

In 2004, it was reported that RGS6 forms a high-affinity complex with DMAP1 and impacts its ability to regulate transcription. RGS6 also coprecipitates with both DMAP1 and Dnmt1 and the interaction between RGS6 and Dnmt1 is DMAP1 dependent.[36] However, the physiological significance of the RGS6–Dnmt1 association remained unknown. Canonically, Dnmt1 functions to maintain genomic DNA methylation patterns in proliferating cells.[97] It also methylates CpG islands in promoter regions, an important mechanism for silencing gene expression.[98] Increasing evidence suggests that Dnmt1-dependent, DNA methylation-mediated silencing of tumor suppressor genes is essential for tumor development and progression, as well as cellular transformation induced by oncogenes, such as Ras.[99–105] Although increased Dnmt1 expression has been observed in a variety of cancers and can occur coincidently within tumors harboring Ras mutations,[106–112] the mechanism underlying overexpression of Dnmt1 in cancers was unclear. In fact, RGS6 facilitates Dnmt1 degradation by scaffolding Dnmt1 and the acetyl transferase Tip60 leading to acetylation-driven Dnmt1 degradation.[28,113] RGS6 impacts Dnmt1 stability, but there is also evidence that Dnmt1 may effect RGS6 as Dnmt1 inhibition in human neural progenitors increases RGS6 expression.[32]

The ability of RGS6 to promote Dnmt1 destabilization is required for suppression of Ras-induced cellular transformation. Similar to results obtained in cells exposed to genotoxic stressors such as Dox, Ras-mediated oncogenic stress triggers RGS6 upregulation. RGS6$^{-/-}$ cells infected with oncogenic Ras exhibit a dramatic increase in anchorage-independent growth and colony formation that can be reversed with inhibitors of Dnmt1 including 5-aza-2′-deoxycytidine (5-aza-dC) used clinically in the treatment of acute myeloid leukemia. Remarkably, in cells lacking RGS6, proteasomal degradation of Dnmt1 is almost completely abolished with RGS6-dependent changes in Dnmt1 expression exactly correlating with alterations in the transcription of Dnmt1 target genes including proapoptotic tumor suppressors *RASSF1A*, *NORE1A*, and *FAS*.[96] Blockade of the caspase cascade increased colony formation in WT but not RGS6$^{-/-}$ MEFS indicating that the proapoptotic actions of RGS6 are required for RGS6-mediated suppression of cellular transformation. Similarly, 5-aza-dC restores the lack of Ras-induced apoptosis induced in cells lacking RGS6.[28] These results support a model in which RGS6 upregulation by Ras blocks Ras-induced transformation via suppression of Dnmt1 protein expression and inhibition of Dnmt1-mediated antiapoptotic activity as well as through the possible direct, proapoptotic actions of RGS6 (Fig. 4).

The RGS domain of RGS6, known only for its GAP activity toward Gα subunits, was sufficient to mediate Tip60 association with RGS6.[28] The crystal structure of the closely related RGS9:Gβ5 complex revealed that the GGL domain is sandwiched between the RGS domain and Gβ5.[114] The observation that DMAP1 binding to the GGL domain of RGS6 is not competitive with Gβ5[36] suggests that the DMAP1 binding surface is between the GGL and RGS domain interface. Thus, Tip60 association with the RGS domain of RGS6 would be on the same surface where we believe Dnmt1 is bound to RGS6, a speculation supported by our finding that RGS6 is required for Tip60-mediated Dnmt1 degradation. Interestingly, the noncanonical RGS protein Axin also associates with Tip60 to mediated p53 acetylation, but this interaction does not occur via Axin's nonfunctional RGS domain.[115] It is possible that RGS6 and Axin bind Tip60 differently in part because Axin possesses an RGS "like" domain that differs structurally from the RGS domain of canonical RGS proteins. The difference in the Tip60 binding site between Axin and RGS6 may be important in determining Tip60 substrate recognition as RGS6 binds Dnmt1 and Axin binds p53 with each of these events facilitating target acetylation.[28,115]

3.3.2 Breast Carcinogenesis

In breast cancer cells and heart, RGS6 promotes apoptosis and mediates the cytotoxic actions of the chemotherapeutic drug Dox.[82,88,96] RGS6 also prevents cellular transformation and induces cell cycle arrest,[28,82] all actions of known tumor suppressor proteins. Predictably, RGS6 also functions as a tumor suppressor in breast *in vivo*.[83] In the DMBA (7,12-dimethylbenza(α) anthracene)-induced tumor model, RGS6 loss results in accelerated tumor initiation and progression. Importantly, even in the absence of a carcinogen, RGS6$^{-/-}$ mice exhibit spontaneous breast tumor formation when aged past 1 year.[83] Just like the pattern observed in human patient samples, RGS6 is expressed in the murine ductal epithelium and lost following chronic DMBA exposure.[82,83] This novel ability of RGS6 to suppress breast tumor formation and growth likely results from a convergence of multiple mechanisms including activation of the DDR, Dnmt1 modulation, cell cycle suppression, and a new role for RGS6 in suppression of breast tumor-specific progrowth receptor signaling cascades.

Both oncogenic activation of progrowth signaling and failure of proapoptotic DNA damage signaling have been implicated as causative factors in DMBA-induced mammary carcinogenesis.[116] *RGS6* deletion cripples carcinogen-induced ROS generation and activation of the DDR (ATM/p53) in the murine ductal epithelium and isolated epithelial cells with a corresponding reduction in apoptosis. As activation of cell death pathways in the face of overwhelming DNA damage is an essential countermeasure against malignant cellular transformation, loss of this pathway likely contributes substantially to the accelerated time course of tumor initiation observed in RGS6$^{-/-}$ mice.[83] Consistent with this supposition, DMBA-induced DNA damage was significantly enhanced in cells lacking RGS6.[83] Though Dnmt1 is upregulated in human breast cancer[117] and polymorphisms in the *DNMT1* gene are associated with breast cancer risk,[118] little is known regarding a functional role for Dnmt1 in breast carcinogenesis. Loss of RGS6 triggers a robust upregulation of Dnmt1 in the mammary epithelium of DMBA-treated mice.[83] Thus, increased methylation-driven repression of proapoptotic tumor suppressor genes may also drive tumorigenesis in the absence of RGS6.

DMBA has been shown to increase expression of cell cycle regulator cyclin D1 with a corresponding decrease in p27^{kip1}, the cyclin-dependent kinase inhibitor.[119] Although basal levels of cyclin D1 and p27^{kip1} in the mammary ductal epithelium did not differ between WT and RGS6$^{-/-}$ mice, increases in cyclin D1 expression and decreases in p27^{kip1} in response to DMBA treatment were significantly enhanced in the ductal epithelium of RGS6 null animals consistent with results obtained in a heterologous expression system.[82,83] In addition, enhanced estrogen receptor (ER)- or human epidermal growth factor receptor 2 (HER2)-mediated growth was observed in mammary epithelial cells lacking RGS6.[83] RGS6-dependent blockade of HER2 signaling is consistent with RGS6-mediated inhibition of Ras, a critical downstream target of HER2 involved in oncogenesis,[120] but, thus far, the mechanism whereby RGS6 blocks ER-dependent growth has yet to be identified. In addition, it remains unclear whether RGS6 directly inhibits Ras or relies on indirect mechanisms (e.g., modulation of Dnmt1 expression) to influence Ras activity. Nevertheless, the ability of RGS6 to block growth downstream of these prominent drivers of breast carcinogenesis is consistent with universal downregulation of RGS6 expression detected in human mammary tumors irrespective of their molecular classification.[83] Instead of modulating a single pathway, RGS6 appears to have the unique capacity to simultaneously suppress the carcinogenic machinery involved in cellular proliferation and promote activation of the tumor-suppressive DDR through mechanisms entirely independent of its canonical G protein regulatory actions.

4. CONCLUSIONS AND FUTURE DIRECTIONS

Herein, we have described a number of phenotypes arising from deletion of the *RGS6* gene or modulation of RGS6 protein levels and summarized current knowledge regarding the mechanism(s) contributing to alterations in human and murine physiology and pathophysiology resulting from perturbations in RGS6 expression or activity. Surprisingly, while known RGS6 domains including the DEP, GGL, and RGS domain have been implicated in these various physiological functions ascribed to RGS6, the ability of RGS6 to interact with and GAP Gα is dispensable for the ability of RGS6 to mediate its antitumor effects. Nevertheless, RGS6 is capable of modulating multiple GPCR signaling cascades influencing motor behavior, mood, and cardiac automaticity. It is important to note that all studies investigating the action of RGS6 *in vivo* to date have relied on global RGS6 knockout mice. The use of such a model system precludes identification of specific tissues or cell types involved in each gross phenotype. In addition, loss of RGS6 activity during development may mask the role of the gene in the adult state or influence phenotypes observed in fully grown animals. Future studies will likely utilize novel gene targeting approaches to begin dissecting further the mechanistic basis for the various phenotypes observed in RGS6$^{-/-}$ mice.

There is a dearth of genetic studies implicating *RGS6* polymorphisms in human disease. Aside from the single report of a polymorphism in the *RGS6* gene associated with increased risk of bladder cancer particularly in smokers[81] and the study identifying RGS6 variants associated with increased heart rate variability,[70] two additional studies have demonstrated a strong linkage between SNPs in the *RGS6* gene and fat intake and adiposity in Hispanics.[121,122] Metabolic parameters have not been evaluated in RGS6$^{-/-}$ mice, but given alterations in diet-induced obesity in mice lacking $G\beta_5$,[14] this may be an interesting future research direction. The lack of genetic variation in the *RGS6* gene linked to human disease is perhaps not that surprising given that changes in RGS6 expression have primarily been observed at the protein level.[82,83] Moving forward, investigations into mechanism(s) controlling RGS6 transcription, translation efficiency, and protein stability might provide valuable insight into the role of RGS6 in various pathologies including cancer and cardiomyopathy. It is important to note that, thus far, RGS6 expression patterns observed in mice closely mirror those in humans[81,83] indicating that the murine model is a viable system to continue investigating the role of RGS6 in human disease.

Due to their role in regulation of GPCR signaling pathways in the nervous system in both physiological and pathophysiological contexts, RGS proteins are emerging as potential therapeutic targets in the treatment of various disorders.[123–125] While GPCR agonists and antagonists have been used successfully in the clinic for years, additional therapeutic benefit could be derived from the additive or singular use of compounds designed to enhance or inhibit RGS protein function. Inhibition of a specific RGS protein might provide additional spatial or temporal selectivity compared with GPCR-targeted drugs, allow for the use of lower agonist doses or a wider therapeutic range, or ameliorate pathological conditions that result directly from RGS protein dysregulation. In addition to their clinical use, such compounds might also have utility in investigating the physiological and pathophysiological functions of RGS proteins, singularly or for multiple proteins, for which there exists no genetic knockout model or those whose loss, alone or in combination, results in embryonic or postnatal lethality. Unlike phenotypes observed in $G\alpha_{i2}$(G148S) mutant mice[126] and $G\beta_5^{-/-}$ mice,[9] RGS6 loss has no detrimental effects on mouse fertility, development, or body weight making RGS6 an attractive novel drug target. In particular, inhibition of RGS6 function or expression would be expected to reduce anxiety and depression, though selective potentiation of 5-HT_{1A}R activation by endogenous ligands,[40,127] ameliorate Dox-induced cardiomyopathy,[96] or treat disorders of the cardiac conduction system.[12,69,70,75] Considering the prevalence of these disorders, such actions could address substantial burdens on the health-care system in the United States and abroad. However, an RGS6 inhibitor would also cause ataxia[40] and increase cancer risk.[28,82,83] The latter would likely preclude the progression of the drug from preclinical to clinical studies in human patients.

Thus, a drug designed to block or destabilize all RGS6 forms (e.g., an inhibitor of the RGS6–$G\beta_5$ interaction) would have limited clinical utility.

It may be possible to eliminate potential off-target effects of RGS6-targeted therapeutics. In particular, the ability of RGS6 to influence mood, motor coordination, and cardiac automaticity are dependent on the GAP activity of RGS6, while the tumor suppressor actions of RGS6 are G protein independent.[82] Therefore, an inhibitor of RGS6's G protein regulatory actions would likely not influence the tumor suppressor actions of RGS6 barring allosteric drug effects. In summation, while the pleiotropic and multifarious nature of RGS6 complicates efforts to elucidate all of the RGS6's functions both G protein dependent and independent, these same aspects of RGS6 physiology could be exploited to design, develop, characterize, and optimize novel RGS6-targeted therapeutics for the treatment of various diseases.

ACKNOWLEDGMENTS

This work was supported by NIH CA161882, AHA 14GRNT20460208, and a University of Iowa Office of the Vice President for Research and Economic Development Award (R.A.F.). A.S. was supported by a Predoctoral Fellowship in Pharmacology/Toxicology from the PhRMA Foundation and a Presidential Fellowship from the University of Iowa Graduate College. B.M. is supported by a Postdoctoral Fellowship from the American Heart Association (14POST20490039).

REFERENCES

1. Hooks SB, Waldo GL, Corbitt J, Bodor ET, Krumins AM, Harden TK. RGS6, RGS7, RGS9, and RGS11 stimulate GTPase activity of Gi family G-proteins with differential selectivity and maximal activity. *J Biol Chem*. 2003;278(12):10087–10093.
2. Jayaraman M, Zhou H, Jia L, Cain MD, Blumer KJ. R9AP and R7BP: traffic cops for the RGS7 family in phototransduction and neuronal GPCR signaling. *Trends Pharmacol Sci*. 2009;30(1):17–24.
3. Grabowska D, Jayaraman M, Kaltenbronn KM, et al. Postnatal induction and localization of R7BP, a membrane-anchoring protein for regulator of G protein signaling 7 family-Gbeta5 complexes in brain. *Neuroscience*. 2008;151(4):969–982.
4. Martemyanov KA, Yoo PJ, Skiba NP, Arshavsky VY. R7BP, a novel neuronal protein interacting with RGS proteins of the R7 family. *J Biol Chem*. 2005;280(7):5133–5136.
5. Drenan RM, Doupnik CA, Jayaraman M, et al. R7BP augments the function of RGS7*Gbeta5 complexes by a plasma membrane-targeting mechanism. *J Biol Chem*. 2006;281(38):28222–28231.
6. Song JH, Waataja JJ, Martemyanov KA. Subcellular targeting of RGS9-2 is controlled by multiple molecular determinants on its membrane anchor, R7BP. *J Biol Chem*. 2006;281(22):15361–15369.
7. Anderson GR, Lujan R, Semenov A, et al. Expression and localization of RGS9-2/G 5/R7BP complex in vivo is set by dynamic control of its constitutive degradation by cellular cysteine proteases. *J Neurosci*. 2007;27(51):14117–14127.
8. Anderson GR, Semenov A, Song JH, Martemyanov KA. The membrane anchor R7BP controls the proteolytic stability of the striatal specific RGS protein, RGS9-2. *J Biol Chem* 2007;282(7):4772–4781.

9. Chen CK, Eversole-Cire P, Zhang H, et al. Instability of GGL domain-containing RGS proteins in mice lacking the G protein beta-subunit Gbeta5. *Proc Natl Acad Sci USA*. 2003;100(11):6604–6609.
10. Krispel CM, Chen CK, Simon MI, Burns ME. Prolonged photoresponses and defective adaptation in rods of Gbeta5 −/− mice. *J Neurosci*. 2003;23(18):6965–6971.
11. Rao A, Dallman R, Henderson S, Chen CK. Gbeta5 is required for normal light responses and morphology of retinal ON-bipolar cells. *J Neurosci*. 2007;27(51):14199–14204.
12. Posokhova E, Wydeven N, Allen KL, Wickman K, Martemyanov KA. RGS6/Gbeta5 complex accelerates IKACh gating kinetics in atrial myocytes and modulates parasympathetic regulation of heart rate. *Circ Res*. 2010;107(11):1350–1354.
13. Xie K, Allen KL, Kourrich S, et al. Gbeta5 recruits R7 RGS proteins to GIRK channels to regulate the timing of neuronal inhibitory signaling. *Nat Neurosci*. 2010;13(6):661–663.
14. Wang Q, Levay K, Chanturiya T, et al. Targeted deletion of one or two copies of the G protein beta subunit Gbeta5 gene has distinct effects on body weight and behavior in mice. *FASEB J*. 2011;25(11):3949–3957.
15. Zhang JH, Pandey M, Seigneur EM, et al. Knockout of G protein beta5 impairs brain development and causes multiple neurologic abnormalities in mice. *J Neurochem*. 2011;119(3):544–554.
16. Xie K, Ge S, Collins VE, et al. Gbeta5-RGS complexes are gatekeepers of hyperactivity involved in control of multiple neurotransmitter systems. *Psychopharmacology*. 2012;219(3):823–834.
17. Xie K, Masuho I, Brand C, Dessauer CW, Martemyanov KA. The complex of G protein regulator RGS9-2 and Gbeta(5) controls sensitization and signaling kinetics of type 5 adenylyl cyclase in the striatum. *Sci Signal*. 2012;5(239):ra63.
18. Anderson GR, Cao Y, Davidson S, et al. R7BP complexes with RGS9-2 and RGS7 in the striatum differentially control motor learning and locomotor responses to cocaine. *Neuropsychopharmacology*. 2010;35(4):1040–1050.
19. Terzi D, Cao Y, Agrimaki I, Martemyanov KA, Zachariou V. R7BP modulates opiate analgesia and tolerance but not withdrawal. *Neuropsychopharmacology*. 2012; 37(4):1005–1012.
20. Zhou H, Chisari M, Raehal KM, et al. GIRK channel modulation by assembly with allosterically regulated RGS proteins. *Proc Natl Acad Sci USA*. 2012; 109(49):19977–19982.
21. Witherow DS, Wang Q, Levay K, et al. Complexes of the G protein subunit gbeta 5 with the regulators of G protein signaling RGS7 and RGS9. Characterization in native tissues and in transfected cells. *J Biol Chem*. 2000;275(32):24872–24880.
22. Snow BE, Betts L, Mangion J, Sondek J, Siderovski DP. Fidelity of G protein beta-subunit association by the G protein gamma-subunit-like domains of RGS6, RGS7, and RGS11. *Proc Natl Acad Sci USA*. 1999;96(11):6489–6494.
23. Posner BA, Gilman AG, Harris BA. Regulators of G protein signaling 6 and 7. Purification of complexes with gbeta5 and assessment of their effects on g protein-mediated signaling pathways. *J Biol Chem*. 1999;274(43):31087–31093.
24. Porter MY, Xie K, Pozharski E, Koelle MR, Martemyanov KA. A conserved protein interaction interface on the type 5 G protein beta subunit controls proteolytic stability and activity of R7 family regulator of G protein signaling proteins. *J Biol Chem*. 2010;285(52):41100–41112.
25. Chatterjee TK, Liu Z, Fisher RA. Human RGS6 gene structure, complex alternative splicing, and role of N terminus and G protein gamma-subunit-like (GGL) domain in subcellular localization of RGS6 splice variants. *J Biol Chem*. 2003; 278(32):30261–30271.

26. Gold SJ, Ni YG, Dohlman HG, Nestler EJ. Regulators of G-protein signaling (RGS) proteins: region-specific expression of nine subtypes in rat brain. *J Neurosci.* 1997;17(20):8024–8037.
27. Mittmann C, Chung CH, Hoppner G, et al. Expression of ten RGS proteins in human myocardium: functional characterization of an upregulation of RGS4 in heart failure. *Cardiovasc Res.* 2002;55(4):778–786.
28. Huang J, Stewart A, Maity B, et al. RGS6 suppresses Ras-induced cellular transformation by facilitating Tip60-mediated Dnmt1 degradation and promoting apoptosis. *Oncogene.* 2014;33(27):3604–3611.
29. Chatterjee TK, Fisher RA. Mild heat and proteotoxic stress promote unique subcellular trafficking and nucleolar accumulation of RGS6 and other RGS proteins. Role of the RGS domain in stress-induced trafficking of RGS proteins. *J Biol Chem.* 2003;278(32):30272–30282.
30. Wang L, Zhou C, Wang Z, et al. Dynamic variation of genes profiles and pathways in the hippocampus of ischemic mice: a genomic study. *Brain Res.* 2011;1372:13–21.
31. Liu Z, Chatterjee TK, Fisher RA. RGS6 interacts with SCG10 and promotes neuronal differentiation. Role of the G gamma subunit-like (GGL) domain of RGS6. *J Biol Chem.* 2002;277(40):37832–37839.
32. Tuggle K, Ali MW, Salazar H, Hooks SB. Regulator of G protein signaling transcript expression in human neural progenitor differentiation: R7 subfamily regulation by DNA methylation. *Neurosignals.* 2014;22(1):43–51.
33. Shin JE, Miller BR, Babetto E, et al. SCG10 is a JNK target in the axonal degeneration pathway. *Proc Natl Acad Sci USA.* 2012;109(52):E3696–E3705.
34. Shin JE, Geisler S, DiAntonio A. Dynamic regulation of SCG10 in regenerating axons after injury. *Exp Neurol.* 2014;252:1–11.
35. Westerlund N, Zdrojewska J, Padzik A, et al. Phosphorylation of SCG10/stathmin-2 determines multipolar stage exit and neuronal migration rate. *Nat Neurosci.* 2011;14(3):305–313.
36. Liu Z, Fisher RA. RGS6 interacts with DMAP1 and DNMT1 and inhibits DMAP1 transcriptional repressor activity. *J Biol Chem.* 2004;279(14):14120–14128.
37. Garzon J, Lopez-Fando A, Sanchez-Blazquez P. The R7 subfamily of RGS proteins assists tachyphylaxis and acute tolerance at mu-opioid receptors. *Neuropsychopharmacology.* 2003;28(11):1983–1990.
38. Sanchez-Blazquez P, Rodriguez-Diaz M, Lopez-Fando A, Rodriguez-Munoz M, Garzon J. The GBeta5 subunit that associates with the R7 subfamily of RGS proteins regulates mu-opioid effects. *Neuropharmacology.* 2003;45(1):82–95.
39. Lopez-Fando A, Rodriguez-Munoz M, Sanchez-Blazquez P, Garzon J. Expression of neural RGS-R7 and Gbeta5 proteins in response to acute and chronic morphine. *Neuropsychopharmacology.* 2005;30(1):99–110.
40. Maity B, Stewart A, Yang J, et al. Regulator of G protein signaling 6 (RGS6) protein ensures coordination of motor movement by modulating GABAB receptor signaling. *J Biol Chem.* 2012;287(7):4972–4981.
41. Rossi DJ, Hamann M, Attwell D. Multiple modes of GABAergic inhibition of rat cerebellar granule cells. *J Physiol.* 2003;548(pt 1):97–110.
42. Drenan RM, Doupnik CA, Boyle MP, et al. Palmitoylation regulates plasma membrane-nuclear shuttling of R7BP, a novel membrane anchor for the RGS7 family. *J Cell Biol.* 2005;169(4):623–633.
43. Jia L, Chisari M, Maktabi MH, et al. A mechanism regulating G protein-coupled receptor signaling that requires cycles of protein palmitoylation and depalmitoylation. *J Biol Chem.* 2014;289(9):6249–6257.
44. Fajardo-Serrano A, Wydeven N, Young D, et al. Association of Rgs7/Gbeta5 complexes with Girk channels and GABAB receptors in hippocampal CA1 pyramidal neurons. *Hippocampus.* 2013;23(12):1231–1245.

45. Ostrovskaya O, Xie K, Masuho I, et al. RGS7/Gbeta5/R7BP complex regulates synaptic plasticity and memory by modulating hippocampal GABABR-GIRK signaling. *eLife*. 2014;3:e02053.
46. Talbot JN, Jutkiewicz EM, Graves SM, et al. RGS inhibition at G(alpha)i2 selectively potentiates 5-HT1A-mediated antidepressant effects. *Proc Natl Acad Sci USA*. 2010;107(24):11086–11091.
47. Riad M, Garcia S, Watkins KC, et al. Somatodendritic localization of 5-HT1A and preterminal axonal localization of 5-HT1B serotonin receptors in adult rat brain. *J Comp Neurol*. 2000;417(2):181–194.
48. Polter AM, Li X. 5-HT1A receptor-regulated signal transduction pathways in brain. *Cell Signal*. 2010;22(10):1406–1412.
49. Stewart A, Maity B, Wunsch AM, et al. Regulator of G-protein signaling 6 (RGS6) promotes anxiety and depression by attenuating serotonin-mediated activation of the 5-HT(1A) receptor-adenylyl cyclase axis. *FASEB J*. 2014;28(4):1735–1744.
50. Pompeiano M, Palacios JM, Mengod G. Distribution and cellular localization of mRNA coding for 5-HT1A receptor in the rat brain: correlation with receptor binding. *J Neurosci*. 1992;12(2):440–453.
51. Richardson-Jones JW, Craige CP, Nguyen TH, et al. Serotonin-1A autoreceptors are necessary and sufficient for the normal formation of circuits underlying innate anxiety. *J Neurosci*. 2011;31(16):6008–6018.
52. Clarke WP, Yocca FD, Maayani S. Lack of 5-hydroxytryptamine1A-mediated inhibition of adenylyl cyclase in dorsal raphe of male and female rats. *J Pharmacol Exp Ther*. 1996;277(3):1259–1266.
53. Beaulieu JM, Zhang X, Rodriguiz RM, et al. Role of GSK3 beta in behavioral abnormalities induced by serotonin deficiency. *Proc Natl Acad Sci USA*. 2008;105(4):1333–1338.
54. Straiker AJ, Borden CR, Sullivan JM. G-protein alpha subunit isoforms couple differentially to receptors that mediate presynaptic inhibition at rat hippocampal synapses. *J Neurosci*. 2002;22(7):2460–2468.
55. Robillard L, Ethier N, Lachance M, Hebert TE. Gbetagamma subunit combinations differentially modulate receptor and effector coupling in vivo. *Cell Signal*. 2000;12(9–10):673–682.
56. Kerchner KR, Clay RL, McCleery G, et al. Differential sensitivity of phosphatidylinositol 3-kinase p110gamma to isoforms of G protein betagamma dimers. *J Biol Chem*. 2004;279(43):44554–44562.
57. Mayeenuddin LH, McIntire WE, Garrison JC. Differential sensitivity of P-Rex1 to isoforms of G protein betagamma dimers. *J Biol Chem*. 2006;281(4):1913–1920.
58. Maurice P, Benleulmi-Chaachoua A, Jockers R. Differential assembly of GPCR signaling complexes determines signaling specificity. *Subcell Biochem*. 2012; 63:225–240.
59. Jaen C, Doupnik CA. RGS3 and RGS4 differentially associate with G protein-coupled receptor-Kir3 channel signaling complexes revealing two modes of RGS modulation. Precoupling and collision coupling. *J Biol Chem*. 2006;281(45):34549–34560.
60. Xie GX, Palmer PP. How regulators of G protein signaling achieve selective regulation. *J Mol Biol*. 2007;366(2):349–365.
61. Abramow-Newerly M, Roy AA, Nunn C, Chidiac P. RGS proteins have a signalling complex: interactions between RGS proteins and GPCRs, effectors, and auxiliary proteins. *Cell Signal*. 2006;18(5):579–591.
62. Ghavami A, Hunt RA, Olsen MA, et al. Differential effects of regulator of G protein signaling (RGS) proteins on serotonin 5-HT1A, 5-HT2A, and dopamine D2 receptor-mediated signaling and adenylyl cyclase activity. *Cell Signal*. 2004;16(6):711–721.
63. Blundell J, Hoang CV, Potts B, Gold SJ, Powell CM. Motor coordination deficits in mice lacking RGS9. *Brain Res*. 2008;1190:78–85.

64. Mark MD, Herlitze S. G-protein mediated gating of inward-rectifier K+ channels. *Eur J Biochem*. 2000;267(19):5830–5836.
65. Kurachi Y. G protein regulation of cardiac muscarinic potassium channel. *Am J Physiol*. 1995;269(4 pt 1):C821–C830.
66. Doupnik CA, Dessauer CW, Slepak VZ, Gilman AG, Davidson N, Lester HA. Time resolved kinetics of direct G(beta 1 gamma 2) interactions with the carboxyl terminus of Kir3.4 inward rectifier K+ channel subunits. *Neuropharmacology*. 1996;35(7):923–931.
67. Doupnik CA, Davidson N, Lester HA, Kofuji P. RGS proteins reconstitute the rapid gating kinetics of gbetagamma-activated inwardly rectifying K+ channels. *Proc Natl Acad Sci USA*. 1997;94(19):10461–10466.
68. Stewart A, Huang J, Fisher RA. RGS proteins in heart: brakes on the vagus. *Front Physiol*. 2012;3:95.
69. Yang J, Huang J, Maity B, et al. RGS6, a modulator of parasympathetic activation in heart. *Circ Res*. 2010;107(11):1345–1349.
70. Posokhova E, Ng D, Opel A, et al. Essential role of the m2R-RGS6-IKACh pathway in controlling intrinsic heart rate variability. *PLoS One*. 2013;8(10):e76973.
71. Chapleau MW, Sabharwal R. Methods of assessing vagus nerve activity and reflexes. *Heart Fail Rev*. 2011;16(2):109–127.
72. Vasan RS, Larson MG, Aragam J, et al. Genome-wide association of echocardiographic dimensions, brachial artery endothelial function and treadmill exercise responses in the Framingham Heart Study. *BMC Med Genet*. 2007;8(suppl 1):S2.
73. Tang YD, Dewland TA, Wencker D, Katz SD. Post-exercise heart rate recovery independently predicts mortality risk in patients with chronic heart failure. *J Card Fail*. 2009;15(10):850–855.
74. Thayer JF, Lane RD. The role of vagal function in the risk for cardiovascular disease and mortality. *Biol Psychol*. 2007;74(2):224–242.
75. Wydeven N, Posokhova E, Xia Z, Martemyanov KA, Wickman K. RGS6, but not RGS4, is the dominant regulator of G protein signaling (RGS) modulator of the parasympathetic regulation of mouse heart rate. *J Biol Chem*. 2014;289(4):2440–2449.
76. Cifelli C, Rose RA, Zhang H, et al. RGS4 regulates parasympathetic signaling and heart rate control in the sinoatrial node. *Circ Res*. 2008;103(5):527–535.
77. Posokhova E, Wydeven N, Allen KL, Wickman K, Martemyanov KA. RGS6/G {beta}5 complex accelerates IKACh gating kinetics in atrial myocytes and modulates parasympathetic regulation of heart rate. *Circ Res*. 2010;107(11):1350–1354.
78. Xie Y, Wolff DW, Wei T, et al. Breast cancer migration and invasion depend on proteasome degradation of regulator of G-protein signaling 4. *Cancer Res*. 2009;69(14):5743–5751.
79. Davydov IV, Varshavsky A. RGS4 is arginylated and degraded by the N-end rule pathway in vitro. *J Biol Chem*. 2000;275(30):22931–22941.
80. Lee MJ, Tasaki T, Moroi K, et al. RGS4 and RGS5 are in vivo substrates of the N-end rule pathway. *Proc Natl Acad Sci USA*. 2005;102(42):15030–15035.
81. Berman DM, Wang Y, Liu Z, et al. A functional polymorphism in RGS6 modulates the risk of bladder cancer. *Cancer Res*. 2004;64(18):6820–6826.
82. Maity B, Yang J, Huang J, Askeland RW, Bera S, Fisher RA. Regulator of G protein signaling 6 (RGS6) induces apoptosis via a mitochondrial-dependent pathway not involving its GTPase-activating protein activity. *J Biol Chem*. 2011;286(2):1409–1419.
83. Maity B, Stewart A, O'Malley Y, Askeland RW, Sugg SL, Fisher RA. Regulator of G protein signaling 6 is a novel suppressor of breast tumor initiation and progression. *Carcinogenesis*. 2013;34(8):1747–1755.
84. Dorsam RT, Gutkind JS. G-protein-coupled receptors and cancer. *Nat Rev Cancer*. 2007;7(2):79–94.

85. Tewey KM, Rowe TC, Yang L, Halligan BD, Liu LF. Adriamycin-induced DNA damage mediated by mammalian DNA topoisomerase II. *Science*. 1984;226(4673):466–468.
86. Lown JW, Sim SK, Majumdar KC, Chang RY. Strand scission of DNA by bound adriamycin and daunorubicin in the presence of reducing agents. *Biochem Biophys Res Commun*. 1977;76(3):705–710.
87. Kurz EU, Douglas P, Lees-Miller SP. Doxorubicin activates ATM-dependent phosphorylation of multiple downstream targets in part through the generation of reactive oxygen species. *J Biol Chem*. 2004;279(51):53272–53281.
88. Huang J, Yang J, Maity B, Mayuzumi D, Fisher RA. Regulator of G protein signaling 6 mediates doxorubicin-induced ATM and p53 activation by a reactive oxygen species-dependent mechanism. *Cancer Res*. 2011;71(20):6310–6319.
89. Guo Z, Kozlov S, Lavin MF, Person MD, Paull TT. ATM activation by oxidative stress. *Science*. 2010;330(6003):517–521.
90. Gianni L, Herman EH, Lipshultz SE, Minotti G, Sarvazyan N, Sawyer DB. Anthracycline cardiotoxicity: from bench to bedside. *J Clin Oncol*. 2008;26(22):3777–3784.
91. Swain SM, Whaley FS, Ewer MS. Congestive heart failure in patients treated with doxorubicin: a retrospective analysis of three trials. *Cancer*. 2003;97(11):2869–2879.
92. Mertens AC, Yasui Y, Neglia JP, et al. Late mortality experience in five-year survivors of childhood and adolescent cancer: the childhood cancer survivor study. *J Clin Oncol*. 2001;19(13):3163–3172.
93. Oeffinger KC, Mertens AC, Sklar CA, et al. Chronic health conditions in adult survivors of childhood cancer. *N Engl J Med*. 2006;355(15):1572–1582.
94. Shizukuda Y, Matoba S, Mian OY, Nguyen T, Hwang PM. Targeted disruption of p53 attenuates doxorubicin-induced cardiac toxicity in mice. *Mol Cell Biochem*. 2005;273(1–2):25–32.
95. Zhao Y, McLaughlin D, Robinson E, et al. Nox2 NADPH oxidase promotes pathologic cardiac remodeling associated with Doxorubicin chemotherapy. *Cancer Res*. 2010;70(22):9287–9297.
96. Yang J, Maity B, Huang J, et al. G-protein inactivator RGS6 mediates myocardial cell apoptosis and cardiomyopathy caused by doxorubicin. *Cancer Res*. 2013;73(6):1662–1667.
97. Bestor TH. The DNA, methyltransferases of mammals. *Hum Mol Genet*. 2000;9(16):2395–2402.
98. Leonhardt H, Cardoso MC. DNA methylation, nuclear structure, gene expression and cancer. *J Cell Biochem Suppl*. 2000;(suppl 35):78–83.
99. Patra SK, Patra A, Zhao H, Dahiya R. DNA methyltransferase and demethylase in human prostate cancer. *Mol Carcinog*. 2002;33(3):163–171.
100. Li LC, Okino ST, Dahiya R. DNA methylation in prostate cancer. *Biochim Biophys Acta*. 2004;1704(2):87–102.
101. Lopez-Serra L, Ballestar E, Fraga MF, Alaminos M, Setien F, Esteller M. A profile of methyl-CpG binding domain protein occupancy of hypermethylated promoter CpG islands of tumor suppressor genes in human cancer. *Cancer Res*. 2006;66(17):8342–8346.
102. Ordway JM, Williams K, Curran T. Transcription repression in oncogenic transformation: common targets of epigenetic repression in cells transformed by Fos, Ras or Dnmt1. *Oncogene*. 2004;23(21):3737–3748.
103. Patra SK. Ras regulation of DNA-methylation and cancer. *Exp Cell Res*. 2008;314(6):1193–1201.
104. Patra SK, Szyf M. DNA methylation-mediated nucleosome dynamics and oncogenic Ras signaling: insights from FAS, FAS ligand and RASSF1A. *FEBS J*. 2008;275(21):5217–5235.

105. Esteller M. Aberrant DNA, methylation as a cancer-inducing mechanism. *Annu Rev Pharmacol Toxicol.* 2005;45:629–656.
106. el-Deiry WS, Nelkin BD, Celano P, et al. High expression of the DNA methyltransferase gene characterizes human neoplastic cells and progression stages of colon cancer. *Proc Natl Acad Sci USA.* 1991;88(8):3470–3474.
107. He S, Wang F, Yang L, et al. Expression of DNMT1 and DNMT3a are regulated by GLI1 in human pancreatic cancer. *PLoS One.* 2011;6(11):e27684.
108. Nakagawa T, Kanai Y, Saito Y, Kitamura T, Kakizoe T, Hirohashi S. Increased DNA methyltransferase 1 protein expression in human transitional cell carcinoma of the bladder. *J Urol.* 2003;170(6 pt 1):2463–2466.
109. Saito Y, Kanai Y, Nakagawa T, et al. Increased protein expression of DNA methyltransferase (DNMT) 1 is significantly correlated with the malignant potential and poor prognosis of human hepatocellular carcinomas. *Int J Cancer.* 2003;105(4):527–532.
110. Etoh T, Kanai Y, Ushijima S, et al. Increased DNA methyltransferase 1 (DNMT1) protein expression correlates significantly with poorer tumor differentiation and frequent DNA hypermethylation of multiple CpG islands in gastric cancers. *Am J Pathol.* 2004;164(2):689–699.
111. Peng DF, Kanai Y, Sawada M, et al. Increased DNA methyltransferase 1 (DNMT1) protein expression in precancerous conditions and ductal carcinomas of the pancreas. *Cancer Sci.* 2005;96(7):403–408.
112. Zhu YM, Huang Q, Lin J, Hu Y, Chen J, Lai MD. Expression of human DNA methyltransferase 1 in colorectal cancer tissues and their corresponding distant normal tissues. *Int J Colorectal Dis.* 2007;22(6):661–666.
113. Du Z, Song J, Wang Y, et al. DNMT1 stability is regulated by proteins coordinating deubiquitination and acetylation-driven ubiquitination. *Sci Signal.* 2010;3(146):ra80.
114. Cheever ML, Snyder JT, Gershburg S, Siderovski DP, Harden TK, Sondek J. Crystal structure of the multifunctional Gbeta5-RGS9 complex. *Nat Struct Mol Biol.* 2008;15(2):155–162.
115. Li Q, Lin S, Wang X, et al. Axin determines cell fate by controlling the p53 activation threshold after DNA damage. *Nat Cell Biol.* 2009;11(9):1128–1134.
116. Vinothini G, Murugan RS, Nagini S. Evaluation of molecular markers in a rat model of mammary carcinogenesis. *Oncol Res.* 2009;17(10):483–493.
117. Agoston AT, Argani P, Yegnasubramanian S, et al. Increased protein stability causes DNA methyltransferase 1 dysregulation in breast cancer. *J Biol Chem.* 2005;280(18):18302–18310.
118. Kullmann K, Deryal M, Ong MF, Schmidt W, Mahlknecht U. DNMT1 genetic polymorphisms affect breast cancer risk in the central European Caucasian population. *Clin Epigenetics.* 2013;5(1):7.
119. Murray SA, Yang S, Demicco E, et al. Increased expression of MDM2, cyclin D1, and p27Kip1 in carcinogen-induced rat mammary tumors. *J Cell Biochem.* 2005;95(5):875–884.
120. Fu X, Osborne CK, Schiff R. Biology and therapeutic potential of PI3K signaling in ER+/HER2-negative breast cancer. *Breast.* 2013;22(suppl 2):S12–S18.
121. Sibbel SP, Talbert ME, Bowden DW, et al. RGS6 variants are associated with dietary fat intake in Hispanics: the IRAS Family Study. *Obesity.* 2011;19(7):1433–1438.
122. Norris JM, Langefeld CD, Talbert ME, et al. Genome-wide association study and follow-up analysis of adiposity traits in Hispanic Americans: the IRAS family study. *Obesity.* 2009;17(10):1932–1941.
123. Sjogren B, Blazer LL, Neubig RR. Regulators of G protein signaling proteins as targets for drug discovery. *Prog Mol Biol Transl Sci.* 2010;91:81–119.
124. Kimple AJ, Bosch DE, Giguere PM, Siderovski DP. Regulators of G-protein signaling and their Galpha substrates: promises and challenges in their use as drug discovery targets. *Pharmacol Rev.* 2011;63(3):728–749.

125. Roman DL, Traynor JR. Regulators of G protein signaling (RGS) proteins as drug targets: modulating G-protein-coupled receptor (GPCR) signal transduction. *J Med Chem*. 2011;54(21):7433–7440.
126. Huang X, Fu Y, Charbeneau RA, et al. Pleiotropic phenotype of a genomic knock-in of an RGS-insensitive G184S Gnai2 allele. *Mol Cell Biol*. 2006;26(18):6870–6879.
127. Stewart A, Maity B, Wunsch AM, et al. Regulator of G-protein signaling 6 (RGS6) promotes anxiety and depression by attenuating serotonin-mediated activation of the 5-HT1A receptor-adenylyl cyclase axis. *FASEB J*. 2014;28(4):1735–1744.

CHAPTER EIGHT

Physiology of RGS10 in Neurons and Immune Cells

Jae-Kyung Lee, Malú G. Tansey[1]
Department of Physiology, Emory University School of Medicine, Atlanta, Georgia, USA
[1]Corresponding author: e-mail address: malu.tansey@emory.edu

Contents

Abstract

Regulator of G protein signaling-10 (RGS10), a GTPase-activating protein (GAP) for $G\alpha_{i3}$, $G\alpha_q$, and $G\alpha_z$, belongs to the D/R12 subfamily based on the homology within the RGS domain and is one of the smallest RGS proteins, outside the RGS box. Although RGS10 lacks the flanking domains or motifs, RGS10 can be targeted to membrane by palmitoylation which markedly increases its net GAP activity. The cAMP-dependent protein kinase A phosphorylates RGS10 on serine 168 (Ser^{168}) which promotes translocation of RGS10 from the cytosol to nucleus. In addition to its GAP function, RGS10 modulates adenylyl cyclase (AC) activity as well as the G protein-gated inwardly rectifying potassium channels. Although the roles of various RGS proteins have been well addressed in previous reviews, the function and mechanism of RGS10 in particular has not been reviewed in-depth. Moreover, recent arrays of studies implicate a role for RGS10 in immune and neuronal cells. RGS10 has been localized to various cell compartments including the cytoplasm and nucleus. In this chapter, we will review its role in neurons and immune cells.

1. INTRODUCTION: THE BASIC STRUCTURE AND ISOFORMS OF RGS10

The regulator of G protein signaling (RGS) family of proteins are GTPase-accelerating proteins (GAPs) that negatively regulate

Progress in Molecular Biology and Translational Science, Volume 133
ISSN 1877-1173
http://dx.doi.org/10.1016/bs.pmbts.2015.01.005

G protein-coupled receptors (GPCR) signaling by increasing the rate of GTP hydrolysis by Gα.[1–3] RGS proteins are highly conserved from yeast to mammals[4] (for a review, see Ref. 5) and are abundant in the retina, brain, heart, and immune organs. RGS proteins differ widely in their sizes and contain a variety of structural domains in addition to the RGS domain and motifs that regulate their activity and determine regulatory binding partners[3,6] (also reviewed in Refs. 7–10).

RGS proteins are divided into six families based on the sequence homology within the RGS domain.[3,6] RGS10 is a 20 kDa protein, one of the smallest RGS family proteins. RGS10 belongs to the D/R12 subfamily and is highly expressed in the brain, thymus, and lymph nodes.[3,11–13] Human RGS10 was identified by its ability to interact with the activated form of $G\alpha_{i3}$ and named RGS10 based on the similarity to a rat gene sequence previously termed RGS10. It contains a 120-amino acid core domain (RGS domain) that is strongly conserved with the yeast Sst2 protein. RGS10 gene is located on human chromosome 10q26.11 and mouse chromosome 7. In humans, there are three splice variants of RGS10, i.e., hRGS10L (181 amino acids (aa)), hRGS10 (173 aa), and hRGS10S (167 aa) (Fig. 1). Interestingly, these variants differ by only a few amino acid residues in the N-terminus as a result of splicing events involving the first three exons.[14] Major types of isoforms are hRGS10 (173 aa) and hRGS10S (167 aa); these two are different in their N-terminus with RGS10S having MEH (aa 1–3) and RGS10 having MQSELCFAD (aa 1–9). Interestingly, the hRGS10 (173 aa) isoform that Hunt *et al.* originally identified possesses the GAP function, but the hRGS10S (167 aa isoform) showed impaired GAP function.[15] Although the RGS domain is essential for the GAP function of RGS proteins, this suggests that the N-terminus of RGS10 is also functionally important to regulate G protein signaling. Moreover, hRGS10S isoform is mainly localized in the nuclear compartment[16] while RGS10 expression has been reported in most of the cellular compartments including membrane, cytoplasm, and nucleus.[17] This suggests that the different RGS10 isoforms may be localized to different compartments inside cells and may play distinct roles.

In mouse, there are two isoforms of RGS10, mRGS10L (181 aa) and mRGS10S (167 aa) that are detected by Western blot at around 21 and 19.5 kDa, respectively. The major isotype that is expressed in mouse immune cells is mRGS10L. Mouse B cells only express mRGS10L but splenic T-cells express both mRGS10L and mRGS10S at low levels. The protein level of RGS10 in rodent brain has been measured using a goat

A

hRGS10L(181aa)

```
MFNRAVSRLS RKRPPSDIHD SDGSSSSSHQ SLKSTAKWAA SLENLLEDPE
GVKRFREFLK KEFSEENVLF WLACEDFKKM QDKTQMQEKA KEIYMTFLSS
KASSQVNVEG QSRLNEKILE EPHPLMFQKL QDQIFNLMKY DSYSRFLKSD
LFLKHKRTEE EEEDLPDAQT AAKRASRIYN T
```

hRGS10(173aa)

```
MQSELCFADI HDSDGSSSSS HQSLKSTAKW AASLENLLED PEGVKRFREF
LKKEFSEENV LFWLACEDFK KMQDKTQMQE KAKEIYMTFL SSKASSQVNV
EGQSRLNEKI LEEPHPLMFQ KLQDQIFNLM KYDSYSRFLK SDLFLKHKRT
EEEEEDLPDA QTAAKRASRI YNT
```

hRGS10S(167aa)

```
MEHIHDSDGS SSSSHQSLKS TAKWAASLEN LLEDPEGVKR FREFLKKEFS
EENVLFWLAC EDFKKMQDKT QMQEKAKEIY MTFLSSKASS QVNVEGQSRL
NEKILEEPHP LMFQKLQDQI FNLMKYDSYS RFLKSDLFLK HKRTEEEEED
LPDAQTAAKR ASRIYNT
```

B

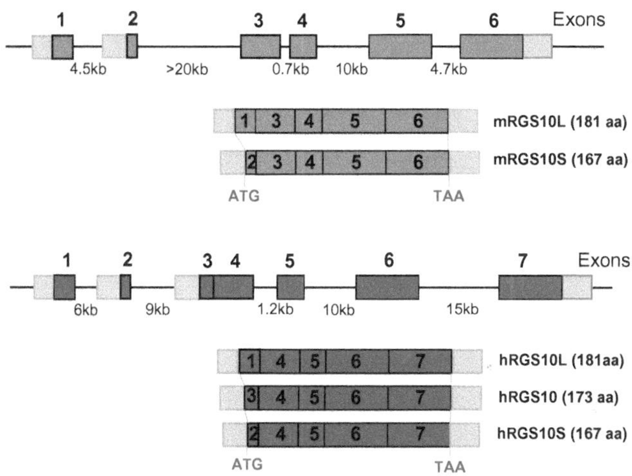

Figure 1 Sequences of human RGS10 isoforms and structures of the mRNA encoded by *Rgs10* gene. (A) Sequence of human RGS10 isoform polypeptides. The sequence data are available from NCBI Web site under accession numbers NM_001005339 (hRGS10L), AF368902 (hRGS10), and NM_002925 (hRGS10S). The "KRAS motif," cAMP-dependent protein kinase A (PKA) phosphorylation sites are in red (or underlined). The differences in the N-terminus between isoforms as results of splicing are in blue. (B) Structure of the mRNA encoded by the mouse and human *RGS10* genes. The mouse *Rgs10* gene is composed of six exons, while the human *Rgs10* gene is composed of seven exons. The differences between the isoforms occur only in the N-terminal sequences as a result of splicing events in the first two exons in mouse or the first three exons in human. (See the color plate.)

anti-RGS10 antibody that was produced against a synthetic peptide identical to 20 residues near the C-terminus of human RGS10. This sequence is more than 90% identical to the corresponding residues in rat and mouse.[17]

Although early evidence suggested that RGS proteins acted primarily as negative regulators of G protein signaling, it has been well documented now that these proteins act as tightly regulated modulators and multifunctional interactors of G protein signaling (reviewed in Ref. 8). In addition, it has been recently appreciated that the non-RGS regions of RGS proteins can provide noncanonical functions distinct from inactivation of Gα subunits or even from G protein signaling entirely (reviewed in Refs. 10,18).

2. REGULATION OF G PROTEIN SIGNALING AND ITS INTERACTING PARTNERS

The main mechanism by which RGS proteins regulate cellular responses is through their GAP activity at Gα subunits, which results in reduced GPCR signaling.[3,13,16,19,20] RGS10 has been shown to associate specifically with the activated forms of $G\alpha_{i3}$, $G\alpha_q$, and $G\alpha_z$ and enhanced their GTPase activities, but did not interact with the $G\alpha_s$ subunit nor affect its GTPase activity.[11] The specific activity accelerating GTP hydrolysis mediated by a conserved RGS domain (120 aa) and its specificity of RGS domain to $G\alpha_i$ is higher than the specificities of RGS domains of other RGS proteins (i.e., RGS4 and GAIP).[21] However, RGS10 is a relatively ineffective GAP in the vesicle-based assay due to its inability to anchor tightly to the vesicle membrane. When the N-terminus domain of RGS10 was replaced with that of RGS4, RGS10 increased its effectiveness as a GAP function by about 1000-fold in the receptor-coupled vesicle-based assay,[22] suggesting membrane localization is necessary for its GAP function. Palmitoylation is a reversible, frequent posttranslational modification in eukaryotic signaling. Palmitoylation of RGS10 at Cys^{66} markedly increases its net GAP activity in vesicle-based assays even though palmitoylation itself inhibited its intrinsic GAP activity.[23]

While some RGS proteins contain a variety of structural domains, such as GGL (G protein gamma-like), DEP (from Dishevelled, EGL-10, and Pleckstrin), or PH (Pleckstrin Homology) domains that regulate their actions and/or enable them to interact with protein-binding partners in addition to the RGS domain,[19] RGS10 does not have any of these domains (Fig. 2).

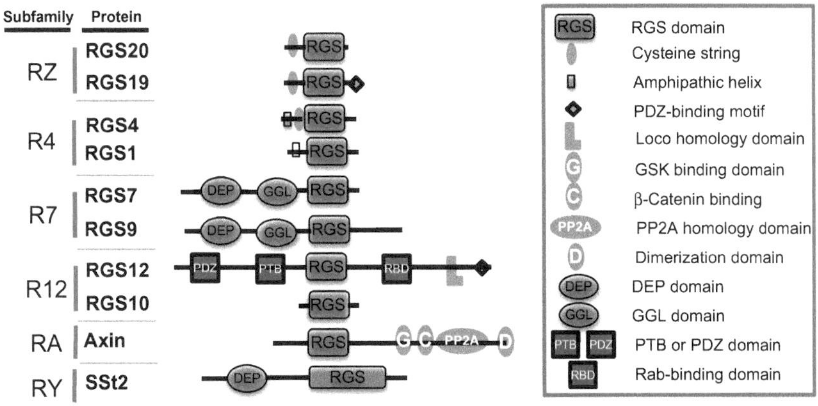

Figure 2 RGS10 lacks the flanking domains and amphipathic helixes. RGS proteins contain a variety of flanking domains, such as GGL (G protein gamma-like), DEP (from Dishevelled, EGL-10, and Pleckstrin), or PH (Pleckstrin Homology) domains that allow them to interact with protein-binding partners.[19] RGS proteins also have amphipathic helixes or cysteine strings that target RGS proteins to the lipid membrane.[13] RGS10 lacks both flanking domains and amphipathic helices.

RGS10 also lacks an amphipathic helix or cysteine string that targets other RGS proteins to the cell lipid membrane compartments. Instead, RGS10 can be targeted to membrane by palmitoylation on the conserved cysteine 66 (Cys^{66}) which markedly increases its net GAP activity.[23] RGS10 contains polypeptide sequences enriched in proline (P), glutamic acid (E), serine (S), and threonine (T) that make up the PEST motif in the C-terminus which likely serves as a proteolytic signal to induce proteasomal degradation. Also, hRGS has been shown to be regulated by cAMP-dependent protein kinase A (PKA) phosphorylation on serine 168 (Ser^{168}) on the "KRAS motif" which is also conserved in mouse orthologues.[24] The phosphorylation promotes translocation of RGS10 from the cytosol to nucleus.[24] These studies strongly suggest that RGS10 protein must become oriented near a membrane environment to display optimal GAP activity in a receptor-coupled system.

There is evidence for receptor-specific and G protein-independent actions of RGS10. It was reported that RGS10 significantly attenuated $G\alpha_i$-dependent signaling mediated by 5-HT1A serotonin but not D2 dopamine receptor-expressing cells.[25] RGS10 modulates adenylyl cyclase (AC) activity by decreasing cAMP accumulation by a G protein-independent pathway.[25] Although the detailed mechanism was not clearly addressed, it was proposed that RGS10 may directly interact with AC or $G\alpha_s$ based on the study that RGS2 required the N-terminus but not the C-terminus, GAP activity, or the RGS box domain for inhibition of AC.[26]

Regulatory effects of RGS10 on G protein-gated inwardly rectifying potassium channels, mediated via its GAP activity toward $G\alpha_i$ proteins, can be regulated by cAMP-dependent kinase (PKA). Burgon and colleagues demonstrated that RGS10 can be phosphorylated at Ser^{168} by PKA which nullifies the RGS10 activity at the plasma membrane. This process controls the G protein-dependent activation of the GIRKs. The attenuation of RGS10 activity was correlated with translocation of its nuclear translocation rather than a reduction of its GAP activity.[24] In addition, Bender and colleagues demonstrated that RGS10 can promote fast activation and deactivation kinetics of GIRK and induce hyperpolarization of excitable cells in response to β_2-adrenergic receptor through a PKA-dependent mechanism. This process also enables a crosstalk between β-adrenergic and muscarinic cholinergic signaling pathway in rat atrial myocytes.[27]

RGS proteins have been shown to be involved in cell proliferation and differentiation.[28,29] The regulation of RGS proteins in $[Ca^{2+}]_i$ oscillations have been reported in immune cells,[30] neural[31] and cardiovascular systems.[32] RGS10 is expressed in osteoclasts and its expression markedly enhanced receptor activator of NF-κB ligand (RANKL)-induced osteoclast differentiation. RGS10 mediates PLCγ activation and $[Ca^{2+}]_i$ oscillation through its $[Ca^{2+}]_i$-dependent dual interaction with Ca^{2+}/calmodulin and PIP_3. The RGS10-mediated intracellular $[Ca^{2+}]_i$ oscillations activate calcineurin and the nuclear factor of activated T-cells, cytoplasmic, calcineurin-dependent 1 (NFATc1) expression for osteoclast terminal differentiation.[33]

The extra-cytoplasmic roles of RGS10, however, remain unclear and underexplored primarily because most GPCRs are localized at the plasma membrane.[34] Recent works from Lee and colleagues reported that nuclear enrichment of RGS10 in microglia in response to inflammatory stimuli enables it to negatively regulate NF-κB, thereby limiting production of microglial-derived inflammatory mediators including soluble tumor necrosis factor (TNF) and lessening the toxic effects on dopaminergic (DA) neurons.[20,35]

3. ROLE OF RGS10 AS A NEGATIVE REGULATOR OF NEUROINFLAMMATION

The RGS proteins were originally identified in tissues including brain, retina, and heart through the observation that G protein signaling proceeded much faster in these tissues than others.[36,37] Proteins of the RGS subfamilies are expressed with some degree of selectivity in the human peripheral tissues and brain.[38] In addition, RGS proteins are differentially regulated in different cell types, tissues, or when cells are in different pathological states. RGS10 is expressed predominantly in brain, especially at high levels in the hippocampus, striatum, and dorsal raphe.[12] RGS10 is present in neurons and is particularly enriched in microglia but not in astrocytes in rat brain[17] and these same expression patterns are found in human brain (unpublished observation).

Microglia are the resident macrophages in the central nervous system (CNS) and are mainly responsible for performing innate immune surveillance in the CNS. Activated microglia change in morphology, increase the expression of genes and release inflammatory mediators, including inducible nitric oxide (NO) synthase and cyclooxygenase-2, NO as well as proinflammatory cytokines including TNF. Although the surveillance properties of microglia are essential for the maintenance of CNS integrity, the excessive, prolonged, or uncontrolled microglial activation has severe and deleterious consequences.[39] Moreover, a wealth of studies have shown that chronic inflammation in the brain is strongly implicated in the pathophysiology of neurodegenerative diseases such as Alzheimer's and Parkinson's disease (PD).

RGS10 is specifically enriched in microglia[17,35] and is the most abundant RGS protein of all the RGS family proteins in microglia where it plays a protective role in the substantia nigra pars compacta (SNpc) against the neurodegenerative effects of chronic peripheral inflammation via regulation of the microglial phenotype.[35,20] RGS10-deficient mice display increased microglial burden in CNS, dysregulated inflammation-related gene expression in microglia and nigral DA neuron degeneration with repeated systemic administration of low-dose lipopolysaccharides (LPS) injection. Specially, proinflammatory cytokines including TNF, IL-1β, and IL-6 secretions are much higher in LPS-stimulated microglia from RGS10 knockout mice compared to wild-type littermate mice.[35] In these studies, RGS10 localizes throughout the cytoplasmic and nuclear compartments during resting condition.[24,35] However, after stimulation with LPS, cytosolic levels of RGS10 are rapidly downregulated while nuclear levels are increased.[35] Consistent with this cytoplasmic to nuclear translocation, a subsequent study demonstrated that RGS10 plays a role as a negative regulator for NF-κB-dependent inflammatory signaling in microglia.[20] However, no evidence of direct interactions between RGS10 and p65 or p50 proteins in the NF-κB complex was found; therefore, the detailed mechanisms by which RGS10 function regulates microglia activation still need to be elucidated. Last, consistent with an important role in this regulation, lentiviral-mediated RGS10 gene therapy targeting glia in the SNpc confers neuroprotective effect on DA neurons against 6-hydroxydopamine (6-OHDA)-induced degeneration.[20] (Fig. 3).

RGS10 expression has been reported in DA neurons although the level is significantly less than that in microglia.[17] DA neurons synthesize dopamine and as such their intracellular environment is inherently one characterized by oxidative stress, a condition that has been proposed to influence their vulnerability to inflammatory stimuli.[40] DA neurons are shown to be extremely sensitive to inflammatory stimuli and oxidative stress.[41–44] Especially, nigral DA neurons are exquisitely sensitive to inflammatory stimuli, including soluble TNF due to their high levels of expression of TNF receptor 1,[41–44] the canonical death receptor.[45] TNF-dependent inflammatory stress has been shown to contribute to the progressive loss of nigral DA neurons in rodent models of parkinsonism based on the studies demonstrating that inhibition of soluble TNF signaling can significantly attenuate endotoxin and oxidative

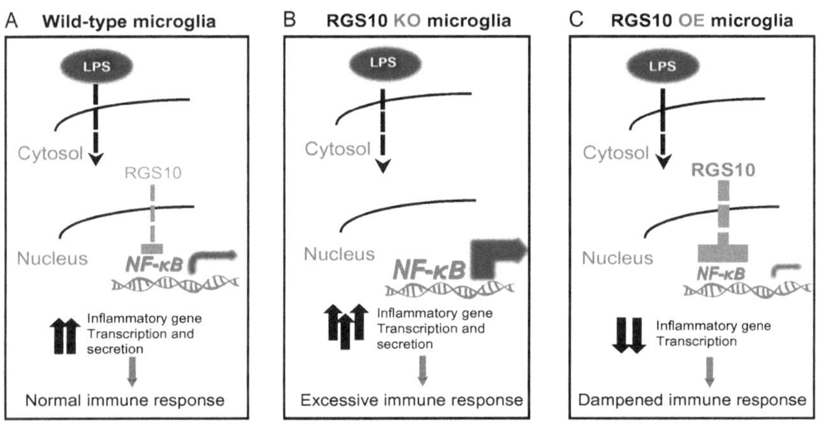

Figure 3 Model for the proposed role of RGS10 as a negative regulator of microglia activation. (A). In microglia, RGS10 negatively regulates NF-κB signaling. When microglia are stimulated by LPS, cytoplasmic RGS10 is turned over which relieves the negative regulation of NF-κB signaling, thereby increasing the transcription of proinflammatory genes. (B). When RGS10 protein levels are significantly downregulated or knocked down, there will be less negative regulation of NF-κB signaling, thereby leading to a heightened inflammatory response upon LPS stimulation. (C). Restoration or overexpression of RGS10 restores negative inhibition of NF-κB signaling which leads to a more normalized response to LPS stimulation in microglia.

neurotoxin-induced degeneration.[46–48] Activation of TNF signaling in neuronally differentiated DA cells result in decreased cellular levels of RGS10 and correlate with TNF-induced cytotoxicity, raising the interesting possibility that loss of neuronal RGS10 may contribute to vulnerability to soluble TNF.[49] To test this, RGS10 was stably overexpressed in neuronally differentiated DA cells where it was found to attenuate TNF-dependent cytotoxicity by inhibiting PARP-1 and caspase 3 cleavage while enhancing PKA/CREB prosurvival signaling.[49] Mechanistically, the biochemical data indicated that a specific serine residue previously shown to be phosphorylated by PKA[24] was critical for mediating the enhanced CREB signaling downstream as RGS10 SA mutant protein cannot mimic the effects of WT RGS10.[49] Consistent with this finding, selective pharmacological inhibition of PKA abolished the neuroprotective effects of RGS10 WT altogether in DA cells treated with TNF.[49] Given that RGS10 can directly interact with PKA as a substrate for phosphorylation at Ser 168, it remains to be elucidated how this interaction contributes to functional modulation of downstream PKA-dependent CREB signaling by RGS10.

In addition to the ability of RGS10 to protect against inflammatory toxicity via regulation of the inflammatory response in microglia, RGS10 has also been demonstrated to protect DA cells against oxidative stress-induced cytotoxicity (6-OHDA) and hydrogen peroxide to cause formation of ROS and RNS (Fig. 4). This suggests that RGS10 regulates prosurvival pathways in DA neurons during toxic insults by virtue of its antiapoptotic and prosurvival actions.

This study raises the possibility that selective modulation of RGS10 levels or bioactivity in neurons may permit neuron-specific potentiation of PKA/CREB-dependent gene transcription to promote neuronal survival without nonselectively activating CREB-dependent proliferation in other cell types. Considering that RGS10 as an important anti-inflammatory modulator in microglia, these novel findings suggest that boosting levels or RGS10 activity in both ventral midbrain microglia and DA neurons may be a novel and doubly effective therapeutic modality to protect DA neurons from chronic inflammatory stress and provide strong rationale for investigating this possibility *in vivo* in preclinical models of PD. Another report indicated that RGS10-deficient mice exhibited impaired osteoclast differentiation and severe osteopetrosis (increased bone density), presumably due to the absence of RGS10-dependent calcium current oscillations and loss of nuclear factor of activated T-cells c1 (NFATc1) expression.[33] Therefore, RGS10 may play an important role in the regulation of intracellular calcium oscillations in

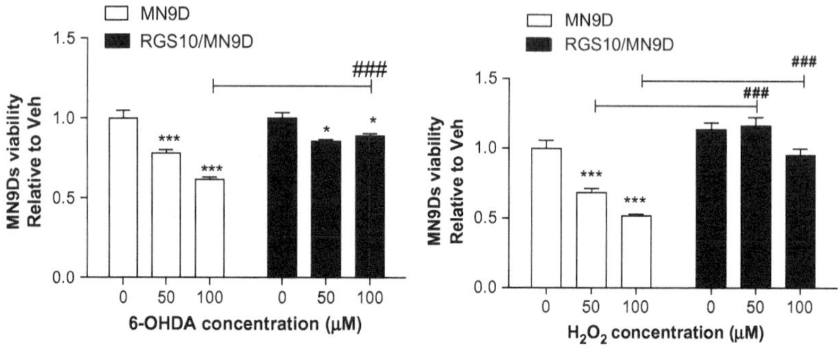

Figure 4 Neuroprotective effect of RGS10 against cytotoxicity induced by 6-OHDA or H_2O_2. Cytotoxicity was examined in ventral mesencephalon dopaminergic neuroblastoma MN9D cells as described previously.[49] 6-Hydroxydopamine-induced or H_2O_2-induced cytotoxicity was measured by the MTS assay in MN9D cells or cell lines stably expressing RGS10. Values represent mean ± SEM. One-way ANOVA followed by Tukey's post hoc test, * and *** denote significant differences between vehicle and treatment within the group at $p < 0.01$ or $p < 0.001$, respectively, ### denotes significant differences between the groups (parental MN9D vs. RGS10 overexpression MN9D cells) at $p < 0.01$ or $p < 0.001$. The values shown are group means ($n = 6$) ± S.E.M (unpublished data).

neurons. Although further studies should be followed to determine the detailed mechanisms, these studies strongly implicate RGS10 as a protective factor in the nigrostriatal pathway against inflammation-induced degeneration. If RGS10 levels decline with age in the CNS, vulnerability for age-related inflammation and neurodegeneration may increase.

4. RGS10 IN PERIPHERAL IMMUNE CELLS

RGS10 is highly expressed in immune organs including the lymph nodes and thymus[14] and is one of the most highly expressed RGS proteins in lymphocytes[50]; yet its function in immune cells remains poorly defined. Although microglia are brain-resident myeloid cells and thus share many similarities in function and expression of many surface markers in common with peripheral macrophages, considerable differences between these cell types have also been reported.[51] Specifically, an elegant fate-mapping study revealed that microglia originate from yolk-sac myeloid progenitors as opposed to the bone-marrow progenitors that give rise to monocytes and eventually peripheral macrophages[52] RGS10 also negatively regulates peripheral macrophage activation (classical, M1) by decreasing expression of genes typically associated with classical activation. RGS10 also exaggerates alternative activation (M2) by enhancing the IL-4-induced alternative activation gene expression profile and blunted phagocytic activity of macrophages.[49] Recently, studies from several laboratories support a critical role of nuclear receptors, PPARs, and PGC-1β in regulating M2 activation.[53–55] Specifically, it has been shown that PPARδ is required for professional phagocytosis in macrophages.[56] Functionally, activation of PPARδ induced the expression of RGS genes 1, 3, 4, 5, 10, 16, and 18 as PPARδ-regulated targets in the macrophage, suggesting modulation of G protein-coupled pathways as a common regulatory mechanism of this receptor[57] and that there may be reciprocal regulation of PPARδ and RGS family protein expression that may contribute to the blunted phagocytic activity of macrophages.

RGS10 also regulates platelet activation. A novel complex comprising the scaffold protein, spinophilin (SPL), the tyrosine phosphatase (SHP-1), and RGS10 or RGS18 in platelets has been reported to modulate platelet activation. Specifically, the SPL/RGS/SHP-1 complexes are present in resting platelets and activation of SHP-1 triggers decay of the SPL/RGS/SHP-1 complexes by sequestering RGS protein and activating platelets.[58] Recently, RGS10 and other RGS mRNA levels in platelets were found to be significantly higher in aspirin-resistant patients than in aspirin-sensitive patients with metabolic syndrome.[59]

RGS10 is also expressed in T-cells where it has been reported to repress the strength of chemokine-dependent T cell adhesion *in vitro* that is mediated by integrin (α4β1).[60] RGS10 attenuates chemokine-upregulated T-cells adhesion medicated by integrin α4β1, and αLβ2. Interestingly, RGS10 mainly regulates the adhesion strength and spreading phase of α4β1-mediated cell attachment. It was suggested that RGS10 contributes to the termination of chemokine-induced signaling and leads to repression of adhesion strengthening mediated by integrin in T-cells.[60]

Last, a role for RGS10 in dendritic cells has been proposed but there is still no detailed information on the mechanisms it regulates. AAV-mediated RGS10 knockdown in periodontal tissues during bacteria-induced inflammatory lesion was reported to attenuate dendritic cell proliferation and maturation which in turn limited T-cell proliferation and maturation and prevented bone resorption and inflammation.[61]

5. RGS10 IN PATHOPHYSIOLOGICAL STATES

The *RGS10* gene is located on human chromosome 10q26.11, a region identified to be linked to age-related maculopathy (ARM). The genotype-identity by descent sharing test showed that the strong association occurred at *rs2039488* which is located between *GRK5* and *RGS10*. However, *RGS10* and *GRK5* expression in ARM showed no clear differences relative to healthy control individuals. However, this negative finding does not necessarily rule out the involvement of these genes in ARM, because the gene expression was determined in the dermal fibroblasts which are not the cell populations that would be expected to have modulation of *RGS10*- and/or *GRK5*-related proteins.[62] Also, an association between the *RGS10* gene and schizophrenia in a Japanese cohort[63] has been reported. However, immunohistochemical analysis of RGS10 protein in postmortem human brain tissue revealed that the levels of RGS10 in schizophrenia and major depression were comparable to that of healthy control subjects.[64,65] Interestingly, RGS proteins do negatively regulate $G_{\alpha i/o}$ protein activity as well as μ-opioid receptor (MORs) function. In support of this possibility, the levels of RGS10 in frontal cortex of short-term opiate abusers are significantly decreased but remain relatively similar to control levels in long-term opiate abusers,[66] suggesting RGS10 may play a role in opiate tolerance, dependence, and/or withdrawal.

REFERENCES

1. Berman DM, Wilkie TM, Gilman AG. GAIP and RGS4 are GTPase-activating proteins for the Gi subfamily of G protein alpha subunits. *Cell.* 1996;86(3):445–452.
2. Siderovski DP, Diverse-Pierluissi M, De Vries L. The GoLoco motif: a Galphai/o binding motif and potential guanine-nucleotide exchange factor. *Trends Biochem Sci.* 1999;24(9):340–341.
3. Ross EM, Wilkie TM. GTPase-activating proteins for heterotrimeric G proteins: regulators of G protein signaling (RGS) and RGS-like proteins. *Annu Rev Biochem.* 2000;69:795–827.
4. Dohlman HG, Thorner J. RGS proteins and signaling by heterotrimeric G proteins. *J Biol Chem.* 1997;272(7):3871–3874.
5. Koelle MR. A new family of G-protein regulators—the RGS proteins. *Curr Opin Cell Biol.* 1997;9(2):143–147.
6. Zheng B, De Vries L, Gist Farquhar M. Divergence of RGS proteins: evidence for the existence of six mammalian RGS subfamilies. *Trends Biochem Sci.* 1999;24(11):411–414.
7. Abramow-Newerly M, Roy AA, Nunn C, Chidiac P. RGS proteins have a signalling complex: interactions between RGS proteins and GPCRs, effectors, and auxiliary proteins. *Cell Signal.* 2006;18(5):579–591.
8. Hollinger S, Hepler JR. Cellular regulation of RGS proteins: modulators and integrators of G protein signaling. *Pharmacol Rev.* 2002;54(3):527–559.
9. Willars GB. Mammalian RGS, proteins: multifunctional regulators of cellular signalling. *Semin Cell Dev Biol.* 2006;17(3):363–376.
10. Kach J, Sethakorn N, Dulin NO. A finer tuning of G-protein signaling through regulated control of RGS proteins. *Am J Physiol Heart Circ Physiol.* 2012;303(1):H19–H35.

11. Hunt TW, Fields TA, Casey PJ, Peralta EG. RGS10 is a selective activator of G alpha i GTPase activity. *Nature*. 1996;383(6596):175–177.
12. Gold SJ, Ni YG, Dohlman HG, Nestler EJ. Regulators of G-protein signaling (RGS) proteins: region-specific expression of nine subtypes in rat brain. *J Neurosci*. 1997;17(20):8024–8037.
13. Sierra DA, Gilbert DJ, Householder D, et al. Evolution of the regulators of G-protein signaling multigene family in mouse and human. *Genomics*. 2002;79(2):177–185.
14. Haller C, Fillatreau S, Hoffmann R, Agenes F. Structure, chromosomal localization and expression of the mouse regulator of G-protein signaling10 gene (mRGS10). *Gene*. 2002;297(1–2):39–49.
15. Ajit SK, Young KH. Analysis of chimeric RGS proteins in yeast for the functional evaluation of protein domains and their potential use in drug target validation. *Cell Signal*. 2005;17(7):817–825.
16. Chatterjee TK, Fisher RA. Cytoplasmic, nuclear, and golgi localization of RGS proteins. Evidence for N-terminal and RGS domain sequences as intracellular targeting motifs. *J Biol Chem*. 2000;275(31):24013–24021.
17. Waugh JL, Lou AC, Eisch AJ, Monteggia LM, Muly EC, Gold SJ. Regional, cellular, and subcellular localization of RGS10 in rodent brain. *J Comp Neurol*. 2005;481(3):299–313.
18. Sethakorn N, Yau DM, Dulin NO. Non-canonical functions of RGS proteins. *Cell Signal*. 2010;22(9):1274–1281.
19. Burchett SA. In through the out door: nuclear localization of the regulators of G protein signaling. *J Neurochem*. 2003;87(3):551–559.
20. Lee JK, Chung J, McAlpine FE, Tansey MG. Regulator of G-protein signaling-10 negatively regulates NF-{kappa}B in microglia and neuroprotects dopaminergic neurons in hemiparkinsonian rats. *J Neurosci*. 2011;31(33):11879–11888.
21. Popov S, Yu K, Kozasa T, Wilkie TM. The regulators of G protein signaling (RGS) domains of RGS4, RGS10, and GAIP retain GTPase activating protein activity in vitro. *Proc Natl Acad Sci USA*. 1997;94(14):7216–7220.
22. Blesch A, Conner JM, Tuszynski MH. Modulation of neuronal survival and axonal growth in vivo by tetracycline-regulated neurotrophin expression. *Gene Ther*. 2001;8(12):954–960.
23. Tu Y, Popov S, Slaughter C, Ross EM. Palmitoylation of a conserved cysteine in the regulator of G protein signaling (RGS) domain modulates the GTPase-activating activity of RGS4 and RGS10. *J Biol Chem*. 1999;274(53):38260–38267.
24. Burgon PG, Lee WL, Nixon AB, Peralta EG, Casey PJ. Phosphorylation and nuclear translocation of a regulator of G protein signaling (RGS10). *J Biol Chem*. 2001;276(35):32828–32834.
25. Ghavami A, Hunt RA, Olsen MA, et al. Differential effects of regulator of G protein signaling (RGS) proteins on serotonin 5-HT1A, 5-HT2A, and dopamine D2 receptor-mediated signaling and adenylyl cyclase activity. *Cell Signal*. 2004;16(6):711–721.
26. Salim S, Sinnarajah S, Kehrl JH, Dessauer CW. Identification of RGS2 and type V adenylyl cyclase interaction sites. *J Biol Chem*. 2003;278(18):15842–15849.
27. Bender K, Nasrollahzadeh P, Timpert M, Liu B, Pott L, Kienitz MC. A role for RGS10 in beta-adrenergic modulation of G-protein-activated K+ (GIRK) channel current in rat atrial myocytes. *J Physiol*. 2008;586(8):2049–2060.
28. Schwable J, Choudhary C, Thiede C, et al. RGS2 is an important target gene of Flt3-ITD mutations in AML and functions in myeloid differentiation and leukemic transformation. *Blood*. 2005;105(5):2107–2114.
29. Appleton CT, James CG, Beier F. Regulator of G-protein signaling (RGS) proteins differentially control chondrocyte differentiation. *J Cell Physiol*. 2006;207(3):735–745.

30. Kehrl JH. Heterotrimeric G, protein signaling: roles in immune function and fine-tuning by RGS proteins. *Immunity*. 1998;8(1):1–10.
31. Sinnarajah S, Dessauer CW, Srikumar D, et al. RGS2 regulates signal transduction in olfactory neurons by attenuating activation of adenylyl cyclase III. *Nature*. 2001;409(6823):1051–1055.
32. Ishii M, Inanobe A, Kurachi Y. PIP3 inhibition of RGS protein and its reversal by Ca2+/calmodulin mediate voltage-dependent control of the G protein cycle in a cardiac K+ channel. *Proc Natl Acad Sci USA*. 2002;99(7):4325–4330.
33. Yang S, Li YP. RGS10-null mutation impairs osteoclast differentiation resulting from the loss of [Ca2+]i oscillation regulation. *Genes Dev*. 2007;21(14):1803–1816.
34. Huang J, Fisher RA. Chapter 5 nuclear trafficking of regulator of g protein signaling proteins and their roles in the nucleus. *Prog Mol Biol Transl Sci*. 2009;86:115–156.
35. Lee JK, McCoy MK, Harms AS, Ruhn KA, Gold SJ, Tansey MG. Regulator of G-protein signaling 10 promotes dopaminergic neuron survival via regulation of the microglial inflammatory response. *J Neurosci*. 2008;28(34):8517–8528.
36. Zerangue N, Jan LY. G-protein signaling: fine-tuning signaling kinetics. *Curr Biol*. 1998;8(9):R313–R316.
37. Arshavsky VY, Pugh Jr EN. Lifetime regulation of G protein-effector complex: emerging importance of RGS proteins. *Neuron*. 1998;20(1):11–14.
38. Larminie C, Murdock P, Walhin JP, et al. Selective expression of regulators of G-protein signaling (RGS) in the human central nervous system. *Brain Res Mol Brain Res*. 2004;122(1):24–34.
39. Lee JK, Tran T, Tansey MG. Neuroinflammation in Parkinson's disease. *J Neuroimmune Pharmacol*. 2009;4:419–429.
40. Tansey MG, Goldberg MS. Neuroinflammation in Parkinson's disease: its role in neuronal death and implications for therapeutic intervention. *Neurobiol Dis*. 2010;37(3):510–518.
41. Aloe L, Fiore M. TNF-alpha expressed in the brain of transgenic mice lowers central tyroxine hydroxylase immunoreactivity and alters grooming behavior. *Neurosci Lett*. 1997;238(1–2):65–68.
42. Carvey PM, Chen EY, Lipton JW, Tong CW, Chang QA, Ling ZD. Intra-parenchymal injection of tumor necrosis factor-alpha and interleukin 1-beta produces dopamine neuron loss in the rat. *J Neural Transm*. 2005;112(5):601–612.
43. Gayle DA, Ling Z, Tong C, Landers T, Lipton JW, Carvey PM. Lipopolysaccharide (LPS)-induced dopamine cell loss in culture: roles of tumor necrosis factor-alpha, interleukin-1beta, and nitric oxide. *Brain Res Dev Brain Res*. 2002;133(1):27–35.
44. McGuire SO, Ling ZD, Lipton JW, Sortwell CE, Collier TJ, Carvey PM. Tumor necrosis factor alpha is toxic to embryonic mesencephalic dopamine neurons. *Exp Neurol*. 2001;169(2):219–230.
45. Tartaglia LA, Ayres TM, Wong GH, Goeddel DV. A novel domain within the 55 kd TNF receptor signals cell death. *Cell*. 1993;74(5):845–853.
46. Harms AS, Barnum CJ, Ruhn KA, et al. Delayed dominant-negative TNF gene therapy halts progressive loss of nigral dopaminergic neurons in a rat model of Parkinson's disease. *Mol Ther*. 2011;19(1):46–52.
47. McCoy MK, Martinez TN, Ruhn KA, et al. Blocking soluble tumor necrosis factor signaling with dominant-negative tumor necrosis factor inhibitor attenuates loss of dopaminergic neurons in models of Parkinson's disease. *J Neurosci*. 2006;26(37):9365–9375.
48. McCoy MK, Ruhn KA, Martinez TN, McAlpine FE, Blesch A, Tansey MG. Intranigral lentiviral delivery of dominant-negative TNF attenuates neurodegeneration and behavioral deficits in hemiparkinsonian rats. *Mol Ther*. 2008;16(9):1572–1579.
49. Lee JK, Chung J, Druey KM, Tansey MG. RGS10 exerts a neuroprotective role through the PKA/c-AMP response-element (CREB) pathway in dopaminergic neuron-like cells. *J Neurochem*. 2012;122:333–343.

50. Moratz C, Harrison K, Kehrl JH. Regulation of chemokine-induced lymphocyte migration by RGS proteins. *Methods Enzymol.* 2004;389:15–32.
51. Guillemin GJ, Brew BJ. Microglia, macrophages, perivascular macrophages, and pericytes: a review of function and identification. *J Leukoc Biol.* 2004;75(3):388–397.
52. Ginhoux F, Greter M, Leboeuf M, et al. Fate mapping analysis reveals that adult microglia derive from primitive macrophages. *Science.* 2010;330(6005):841–845.
53. Vats D, Mukundan L, Odegaard JI, et al. Oxidative metabolism and PGC-1beta attenuate macrophage-mediated inflammation. *Cell Metab.* 2006;4(1):13–24.
54. Bouhlel MA, Derudas B, Rigamonti E, et al. PPARgamma activation primes human monocytes into alternative M2 macrophages with anti-inflammatory properties. *Cell Metab.* 2007;6(2):137–143.
55. Kang K, Reilly SM, Karabacak V, et al. Adipocyte-derived Th2 cytokines and myeloid PPARdelta regulate macrophage polarization and insulin sensitivity. *Cell Metab.* 2008;7(6):485–495.
56. Mukundan L, Odegaard JI, Morel CR, et al. PPAR-delta senses and orchestrates clearance of apoptotic cells to promote tolerance. *Nat Med.* 2009;15(11):1266–1272.
57. Barish GD, Atkins AR, Downes M, et al. PPARdelta regulates multiple proinflammatory pathways to suppress atherosclerosis. *Proc Natl Acad Sci USA.* 2008;105(11):4271–4276.
58. Ma P, Cierniewska A, Signarvic R, et al. A newly identified complex of spinophilin and the tyrosine phosphatase, SHP-1, modulates platelet activation by regulating G protein-dependent signaling. *Blood.* 2012;119(8):1935–1945.
59. Mao Y, Lei L, Su J, Yu Y, Liu Z, Huo Y. Regulators of G protein signaling are up-regulated in aspirin-resistant platelets from patients with metabolic syndrome. *Pharmazie.* 2014;69(5):371–373.
60. Garcia-Bernal D, Dios-Esponera A, Sotillo-Mallo E, Garcia-Verdugo R, Arellano-Sanchez N, Teixido J. RGS10 restricts upregulation by chemokines of T cell adhesion mediated by alpha4beta1 and alphaLbeta2 integrins. *J Immunol.* 2011;187(3):1264–1272.
61. Yang S, Hao L, McConnell M, et al. Inhibition of Rgs10 expression prevents immune cell infiltration in bacteria-induced inflammatory lesions and osteoclast-mediated bone destruction. *Bone Res.* 2013;1(3):267–281.
62. Jakobsdottir J, Conley YP, Weeks DE, Mah TS, Ferrell RE, Gorin MB. Susceptibility genes for age-related maculopathy on chromosome 10q26. *Am J Hum Genet.* 2005;77(3):389–407.
63. Hishimoto A, Shirakawa O, Nishiguchi N, et al. Novel missense polymorphism in the regulator of G-protein signaling 10 gene: analysis of association with schizophrenia. *Psychiatry Clin Neurosci.* 2004;58(5):579–581.
64. Rivero G, Gabilondo AM, Garcia-Sevilla JA, et al. Brain RGS4 and RGS10 protein expression in schizophrenia and depression. Effect of drug treatment. *Psychopharmacology (Berl).* 2013;226(1):177–188.
65. Rivero G, Gabilondo AM, Garcia-Sevilla JA, La Harpe R, Morentin B, Javier Meana J. Characterization of regulators of G-protein signaling RGS4 and RGS10 proteins in the postmortem human brain. *Neurochem Int.* 2010;57(7):722–729.
66. Rivero G, Gabilondo AM, Garcia-Fuster MJ, La Harpe R, Garcia-Sevilla JA, Meana JJ. Differential regulation of RGS proteins in the prefrontal cortex of short- and long-term human opiate abusers. *Neuropharmacology.* 2012;62(2):1044–1051.

CHAPTER NINE

Regulator of G Protein Signaling 14: A Molecular Brake on Synaptic Plasticity Linked to Learning and Memory

Paul R. Evans*, Serena M. Dudek†, John R. Hepler*[,1]
*Department of Pharmacology, Emory University School of Medicine, Rollins Research Center, Atlanta, Georgia, USA
†Neurobiology Laboratory, National Institute of Environmental Health Sciences, National Institutes of Health, Research Triangle Park, North Carolina, USA
[1]Corresponding author: e-mail address: jhepler@emory.edu

Contents

Abstract

The regulators of G protein signaling (RGS) proteins are a diverse family of proteins that function as central components of G protein and other signaling pathways. In the brain, regulator of G protein signaling 14 (RGS14) is enriched in neurons in the hippocampus where the mRNA and protein are highly expressed. This brain region

Progress in Molecular Biology and Translational Science, Volume 133
ISSN 1877-1173
http://dx.doi.org/10.1016/bs.pmbts.2015.03.006

plays a major role in processing learning and forming new memories. RGS14 is an unusual RGS protein that acts as a multifunctional scaffolding protein to integrate signaling events and pathways essential for synaptic plasticity, including conventional and unconventional G protein signaling, mitogen-activated protein kinase, and, possibly, calcium signaling pathways. Within the hippocampus of primates and rodents, RGS14 is predominantly found in the enigmatic CA2 subfield. Principal neurons within the CA2 subfield differ from neighboring hippocampal regions in that they lack a capacity for long-term potentiation (LTP) of synaptic transmission, which is widely viewed as the cellular substrate of learning and memory formation. RGS14 was recently identified as a natural suppressor of LTP in hippocampal CA2 neurons as well as forms of learning and memory that depend on the hippocampus. Although CA2 has only recently been studied, compelling recent evidence implicates area CA2 as a critical component of hippocampus circuitry with functional roles in mediating certain types of learning and memory. This review will highlight the known functions of RGS14 in cell signaling and hippocampus physiology, and discuss potential roles for RGS14 in human cognition and disease.

1. INTRODUCTION

The regulators of G protein signaling (RGS) family are central regulatory components of heterotrimeric G protein signaling pathways, and emerging evidence has revealed unexpected complexity and noncanonical roles for RGS proteins.[1–5] RGS proteins interact directly with activated Gα-GTP subunits and function as GTPase-accelerating proteins (GAPs) to limit the lifetime of G protein signaling events. Our understanding of RGS protein biology has evolved to include an appreciation for additional functions of complex RGS proteins that contain multiple protein-interacting domains. RGS14 is a highly unusual RGS protein with a multidomain structure that allows it to interact with binding partners from multiple signaling pathways.[6] In brain, particularly in hippocampus, neurons utilize these biochemical cascades to process learning input and encode memories by modulating the strength of synaptic transmission at particular synapses—a process generally referred to as synaptic plasticity. We recently identified RGS14 as a natural suppressor of synaptic plasticity in hippocampal CA2 neurons, as well as hippocampal-based learning and memory.[7] This review will focus on the known signaling functions of RGS14, its roles in regulating hippocampal physiology, and the implications for human behavior and disease.

2. MOLECULAR CHARACTERIZATION AND THE CELLULAR FUNCTIONS, LOCALIZATION, AND REGULATION OF RGS14

2.1 Overview of G Protein/GPCR/RGS Signaling

Established models propose that agonist binding to a G protein-coupled receptor (GPCR) induces a conformational change in the receptor, which then activates heterotrimeric G proteins (Gαβγ) by acting as a guanine nucleotide exchange factor (GEF) to catalyze the exchange of GTP for GDP on the Gα subunit. The activated Gα-GTP dissociates from Gβγ, and both subunits are free to interact with downstream effectors.[8–11] RGS proteins recognize and directly bind to activated Gα-GTP subunits through a conserved RGS domain and act as GAPs by stimulating the intrinsic GTPase activity of the Gα subunit to hydrolyze GTP to GDP, thereby deactivating G protein signaling. Compelling evidence now indicates that GPCRs are platforms where specific sets of proteins assemble to execute receptor-specific signaling events. RGS proteins are central components of GPCR signaling complexes that fine-tune G protein signaling and serve as multifunctional integrators of these pathways.[2,5,12–14]

The RGS protein family consists of almost 40 members that share a conserved RGS domain, which selectively binds to activated Gα-GTP subunits and, in nearly all cases, confers GAP activity.[1–4,12] The RGS proteins are categorized into subfamilies based on sequence homology of the RGS domains and GAP function.[1–3] The structures, functions, and regions flanking the RGS domains vary widely among the diverse members of this protein family ranging from simple polypeptides comprised of only an RGS domain with limited flanking regions to larger, more complex members with additional protein-binding domains. The presence of binding domains for additional proteins allows these complex RGS proteins to serve as multifunctional integrators of G protein signaling pathways that regulate cell signaling and organ physiology.[2,4] This review will highlight the established signaling roles of RGS14, and how its capacity to bridge separate signaling networks could affect the acquisition of learning leading to episodic memory.

2.2 Structure, Signaling, and Regulation of RGS14

RGS14 is a 61-kDa protein classified in the D/R12 subfamily of RGS proteins along with its closest related RGS proteins, RGS10 and RGS12.[3,12,15]

RGS10 is a small protein (~20 kDa) that only shares high sequence identity with the RGS domain of RGS14. By contrast, RGS14 and RGS12 are more complex proteins with multidomain structures.[15,16] RGS14 contains an N-terminal RGS domain, which binds to and exerts GAP activity toward activated Gαi/o-GTP subunits to limit the duration of heterotrimeric G protein signaling.[17] In addition, RGS14 also contains two tandem Ras/Rap-binding domains (RBDs) and a G protein regulatory (GPR, also referred to as a GoLoco motif) motif (Fig. 1).[16–18] RGS14 preferentially binds activated H-Ras-GTP through its first RBD,[19–21] and the tandem RBD region also mediates interactions with Rap2-GTP and Raf kinases.[16,20] We recently have found that this region of RGS14 also can bind calmodulin (CaM) in a Ca^{2+}-dependent manner.[22] RGS14 selectively binds inactive Gαi1/3-GDP through the GPR motif to inhibit guanine nucleotide exchange (i.e., GDI activity) and localize to cellular plasma membranes.[23–26] The presence of these multiple protein-binding domains indicates that RGS14 serves other important signaling functions in addition to the canonical GAP activity of RGS proteins. Of note, all of RGS14's identified binding partners have key roles in synaptic plasticity in the hippocampus making RGS14 well positioned to modulate neuronal physiology. We will first review the known signaling functions, localization, and regulation of RGS14 so as to provide context for our subsequent discussion of RGS14's roles in brain, and our proposed model for how RGS14 integrates signaling in hippocampal CA2 neurons. We also will highlight recent findings about newly appreciated roles for hippocampal area CA2, the brain region where RGS14 is expressed.

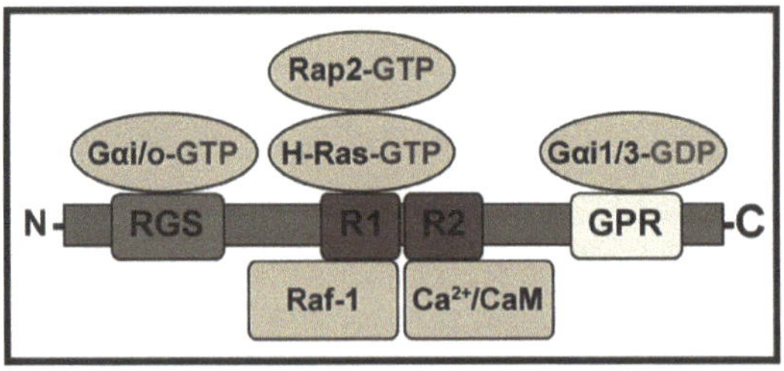

Figure 1 RGS14 domain structure and identified binding partners. RGS14 directly binds activated Gαo and Gαi subunits through its N-terminal RGS domain and selectively interacts with inactive Gαi1 and Gαi3 at the GPR motif. RGS14 also contains tandem Ras/Rap-binding domains (R1 and R2) that directly bind activated H-Ras, Rap2, Ca^{2+}/CaM, and Raf kinases. (See the color plate.)

RGS14 and RGS12 are the only RGS proteins that contain two Gα-binding domains, the GPR motif and the conserved RGS domain, allowing these members to interact with both active GTP-bound and inactive GDP-bound Gα subunits.[6,17,23] RGS14 can clearly operate as a conventional GAP for members of the Gαi/o subfamily,[6,17,18] and we propose that RGS14 first engages Gαi-GTP following activation of a GPCR. In support of this model, native RGS14 is detected in both the cytosolic and membrane fractions of brain extracts[17] and is visible in dendritic shafts and spine heads of hippocampal neurons by electron microscopy, but not at the plasma membrane.[7] A small fraction of RGS14 may be recruited from the cytosol to act as a conventional GAP and limit Gαi-GTP signaling, and then become captured at the plasma membrane by the resulting Gαi-GDP via its GPR motif. Once at the membrane, the RGS14:Gαi1-GDP complex can serve as a signaling node to engage other signaling proteins/pathways. Curiously, truncated forms of RGS14 lacking the RGS domain can bind and engage other smaller RGS protein such as RGS2 and RGS4 to markedly enhance their GAP activity,[27,28] and this interaction is facilitated by the tandem Ras-binding domains (RBDs, R1 and R2).[27] The physiological significance of this unexpected interaction is unclear, but splice variants of RGS14 lacking the RGS domain have been reported that may operate in this manner.[27]

The Gαi-GDP-interacting GPR motif found in RGS14 is a defining feature shared by all members of the Group II activator of G protein signaling (AGS) proteins family.[29] Recent evidence indicates these AGS proteins participate in "unconventional" G protein signaling that regulates cell and organ physiology.[30,31] In contrast to traditional models of heterotrimeric G protein signaling initiated by GPCR activation, unconventional G protein signaling models[6,29–32] posit that the Gα protein exists in a resting state complex with a GPR protein, in place of Gβγ, until activated by either a GPCR or a nonreceptor GEF. The presence of these two Gα-binding sites uniquely positions RGS14 and RGS12 at the interface of these two protein (RGS and AGS) families to serve functions distinct from other conventional RGS proteins.[6]

In this regard, we and others have shown that RGS14 utilizes its complex protein architecture to integrate conventional and unconventional G protein signaling pathways.[26,33,34] Cell imaging studies have shown that RGS14 localization in cells is heavily influenced by interactions with G proteins. When expressed in cells, recombinant RGS14 alone predominantly localizes to the cytosol, but it rapidly moves between the cytosol and nucleus and associates with centrosomes in a cell cycle-dependent manner.[25,34] Coexpression of Gαi1/3 recruits RGS14 from the cytosol to the plasma membrane where they colocalize.[18,25,26] Deletion or selective inactivation of a nuclear localization sequence within the GPR motif eliminates the capacity of RGS14 to bind Gαi1/3 at the GPR motif, recruit RGS14 to the plasma membrane, or localize to the nucleus, indicating G protein interactions with the GPR motif are necessary for proper subcellular localization of RGS14.[23–25] Possible roles for RGS14 in the nucleus remain a mystery at this time. Native RGS14 has not yet been observed in the nuclei of hippocampal neurons,[7] but this localization may be triggered by specific signaling events or may be relevant in B- and T-cell lymphocytes where RGS14 also is highly expressed. A role for RGS14 in the nuclei of hippocampal neurons cannot be ruled out and remains a topic of exploration.

Evidence suggests that RGS14 functionally interacts with newly appreciated members of unconventional G protein pathways and participates in receptor-independent G protein signaling.[26] Resistance to inhibitors of cholinesterase-8A (Ric-8A, also referred to as synembryn) is a cytosolic GEF that binds to and catalyzes nucleotide exchange on inactive Gαi-GDP subunits[35] and regulates other GPR-containing proteins in complex with Gαi1-GDP.[26,36,37] When RGS14 and Ric-8A were transfected into cells, alone or together, both proteins colocalize in the cytosol. Coexpression of wild-type Gαi1 causes both proteins to translocate to the plasma membrane suggesting they may form a functional signaling complex. Ric-8A interacts with RGS14 through the tandem RBD region to induce dissociation of the RGS14:Gαi1-GDP complex in cells and *in vitro* by competing with RGS14 for Gαi1 to exert GEF activity. A role for RGS14 in unconventional G protein signaling could be physiologically relevant since RGS14 and Ric-8A natively colocalize in hippocampal CA2 neurons,[26] but physiological roles for RGS14:Gαi1 and Ric-8A interactions remain undefined.

The RGS14:Gαi1 signaling complex can also functionally interact with a Gαi-linked GPCR, thereby integrating conventional and unconventional G protein pathways.[33] Bioluminescence resonance energy transfer (BRET) studies in live cells validated that RGS14 binds Gαi1/3 at the plasma membrane through the GPR motif, and showed that RGS14 selectively forms a complex with the Gi/o-linked α_{2A}-adrenergic receptor (α_{2A}-AR) in a Gαi/o-dependent manner via the GPR motif. Interestingly, agonist stimulation of the α_{2A}-AR markedly decreased its interaction with RGS14 with no effect on the RGS14:Gαi1 complex. Certain other Group II AGS proteins also interact with Gαi1 and associate with Gi/o-linked

GPCRs,[38,39] but these proteins seem to dissociate from Gαi1 following agonist stimulation, suggesting that the RGS14:Gαi1-GDP complex may be regulated and function differently from other AGS:Gαi1 complexes. Ric-8A induced dissociation of Gαi1 from RGS14 and α_{2A}-AR, and even more so following agonist stimulation. These studies demonstrated that RGS14 functionally integrates conventional and unconventional G protein signaling pathways in live cells and is regulated in a manner that is distinct from other RGS- and GPR-containing proteins.[33]

Besides having two Gα-binding domains, RGS14 also contains tandem RBDs that position it to integrate G protein and mitogen-activated protein kinase (MAPK) cascades.[20,21] RGS14 can suppress extracellular signal-regulated kinase (ERK) activation by a Gαi1- and H-Ras/Raf-dependent mechanism.[20] RGS14 interacts selectively with activated H-Ras-GTP and Raf kinases simultaneously to form a ternary complex, and these partners facilitate each other's interactions with RGS14. RGS14 binds activated H-Ras and Raf kinases in cells to inhibit platelet-derived growth factor-stimulated ERK 1/2 phosphorylation. When a targeted loss-of-function mutation that prevents H-Ras binding was introduced into the R1 domain of RGS14 (R333L), RGS14 could no longer suppress ERK phosphorylation indicating that RGS14 directly binds H-Ras-GTP to negatively regulate ERK signaling. Coexpression of Gαi1 also prevented RGS14 from regulating ERK by disrupting its interactions with Raf kinases, but not H-Ras. Taken together, these data indicate that (1) RGS14 interactions with Gαi1 and Raf kinases are mutually exclusive, and (2) RGS14 interactions with H-Ras and Raf kinases are necessary for RGS14 to inhibit ERK phosphorylation. These results suggest that RGS14 may serve as a G protein-regulated molecular switch to modulate H-Ras/ERK signaling depending upon the upstream signal and proteins in complex.

Following up on these findings, BRET studies examining Gαi1:RGS14:H-Ras interactions in live cells support a model in which RGS14 toggles between G protein and MAPK signaling pathways.[21] These findings confirmed that RGS14 binds activated H-Ras-GTP through the first RBD,[19,20] and binding of inactive Gαi1-GDP enhances RGS14 interactions with H-Ras, likely promoting the assembly of a trimeric Gαi1:RGS14:H-Ras complex.[21] Cell imaging data revealed that coexpression of Gαi1 and/or constitutively active H-Ras (G12V) recruits RGS14 from the cytosol to plasma membrane where they colocalize. Consistent with this idea, activated H-Ras membrane localization is required for RGS14/H-Ras interactions.

In addition to Gαi1-mediated regulation of H-Ras:RGS14 binding, specific GPCRs also can regulate the RGS14/H-Ras complex. As mentioned earlier, RGS14 associates with the Gi/o-linked α_{2A}-AR only in the presence of Gαi/o proteins.[33] Coexpression of the α_{2A}-AR and Gαi1 did not affect the basal interactions between RGS14 and active H-Ras, but agonist stimulation of the α_{2A}-AR resulted in a decreased RGS14/H-Ras BRET signal. The RGS14:Gαi1 complex association with α_{2A}-AR was partially blocked in the presence of activated H-Ras, but H-Ras also reduced the agonist-induced dissociation of the RGS14/α_{2A}-AR complex observed only in the presence of Gαi1. These results demonstrate that H-Ras and the α_{2A}-AR reciprocally regulate one another's association with RGS14 in a Gαi1-dependent manner. Regulation of the RGS14/H-Ras complex by α_{2A}-AR activation could induce a new conformation in RGS14 allowing it to engage downstream effectors. For example, the complex might rearrange to position RGS14 to GAP the Gαi/o subunit activated by the α_{2A}-AR. Another possibility is that the Gαi1:RGS14:H-Ras complex interacts with distinct effector(s) that specifically recognize the ternary complex. It remains to be determined whether this complex is subject to similar regulation by other GPCRs or nonreceptor GEFs, such as Ric-8A.

The structural basis for how RGS14 interacts with its partners to integrate these signaling pathways remains undefined at this time, but is an active area of research. The evidence discussed so far indicates that RGS14 undergoes significant intramolecular conformational rearrangements, depending on the binding partners, consistent with its role as a scaffolding protein. In particular, Gαi1 interactions with RGS14 promote association with activated H-Ras[21] and/or the α_{2A}-AR,[33] whereas G protein binding inhibits binding to Raf kinases.[20] Although the structures of individual domains contained in RGS14 have been elucidated, structural data of the full-length protein could yield great insight into the molecular basis of the tightly regulated assembly of these signaling complexes. Ongoing experiments are examining whether RGS14 can bind multiple Gα proteins simultaneously through the RGS domain and the GPR motif, and how this affects the overall structure and function of RGS14.

RGS14 is also subject to posttranslational modifications that play an important role in modulating its functions. Kinases are prominent downstream targets of the cellular pathways linked to RGS14, and protein phosphorylation regulates the activity of many proteins, including RGS proteins.[40] RGS14 negatively regulates Gαi/o-GTP signaling through the GAP activity of its RGS domain as well as GDI activity at the GPR motif, and Gαi/o activation canonically inhibits adenylyl cyclase, thereby reducing cellular levels of cAMP and, consequently, cAMP-dependent protein kinase A (PKA) activity. We have previously shown that RGS14 is phosphorylated at two sites (Ser 258 and Thr 494) by PKA *in vitro* and natively in B35 neuroblastoma cells.[41] PKA-mediated phosphorylation of RGS14 at T494 increases its affinity for Gαi1-GDP binding at the GPR motif *in vitro*, suggesting this modification could promote and stabilize the assembly of a RGS14:Gαi1 signaling complex in cells. Though speculative, PKA-directed phosphorylation of RGS14 could serve as a feed-forward mechanism to potentiate PKA activity by facilitating interactions with Gαi1, which recruits RGS14 to the plasma membrane and is required for RGS14 to associate with a GPCR or cytosolic GEF. Once at the plasma membrane, RGS14's GAP activity could limit Gαi/o-GTP signaling following GPCR activation (or receptor-independent GEFs) to alleviate inhibition of adenylyl cyclase and elevate cAMP/PKA activity. PKA phosphorylation could also prolong the lifetime of G$\beta\gamma$ signaling events by stabilizing RGS14:Gαi1-GDP complex formation. Implicit to this model is the idea that PKA phosphorylation promotes an RGS14:Gαi1 complex at the plasma membrane to nucleate association with other binding partners in a multiprotein signaling complex (G protein, H-Ras-GTP, Ca^{2+}/CaM) to regulate specific downstream effector pathways.

RGS14 also is phosphorylated by ERK at serine 52 *in vitro*.[40] Whether this modification occurs in cells and the functional effects of this event are currently unknown, but remain a topic of interest. The fact that RGS14 directly engages the H-Ras/ERK signaling pathway suggests some feedback regulation. Proteomics studies have identified several additional phosphorylated residues on RGS14 from rodent and human tissues.[42] Therefore, it is likely that RGS14 is phosphorylated by many more kinases in its native environment. Identifying the kinases and cellular signals that trigger phosphorylation of RGS14 and the functional consequences of these modifications will provide great insight into the complex regulation of RGS14 *in vivo*. RGS14 may also be subject to other posttranslational modifications (e.g., ubiquitination, lipid modifications) that could strongly influence its cellular functions.

In summary, RGS14 is a complex and highly unusual RGS protein that functionally integrates conventional G protein signaling with unconventional G protein pathways, MAPK signaling pathways, and possibly calcium signaling pathways. The signaling functions, subcellular localization, and binding partner interactions of RGS14 are tightly regulated by protein interactions and posttranslational modifications. Therefore, RGS14 uniquely sits at the interface between multiple signaling networks that have well-defined roles in physiology, and especially so in the regulation of synaptic plasticity within the hippocampal neurons where RGS14 is highly expressed.

3. RGS14 NATURALLY LIMITS LEARNING AND SYNAPTIC PLASTICITY IN HIPPOCAMPAL CA2

Previous studies characterizing the tissue distribution of RGS14 found that this signaling protein is present in brain, spleen, and B and T lymphocytes.[15–18,43] Native RGS14 is found at high levels in brain of various species including mouse, rat, nonhuman primate, and human.[7,15,17,43–46] In adult rodents, RGS14 is most highly expressed in brain, specifically in neurons.[7,17,43] Most recently, we have shown that RGS14 mRNA/protein expression is upregulated during postnatal mouse brain development reaching its highest levels in adulthood.[43] Within the adult mouse brain, we demonstrated that RGS14 protein is highly enriched in spines and dendrites of pyramidal neurons in the CA2 subfield of the hippocampus.[7,43] Although RGS14 likely has important functions in immune cells, this review will focus on the defined role of RGS14 in episodic learning and memory and hippocampus function.

3.1 Episodic Memory Formation Depends on the Hippocampus

RGS14 is highly expressed and restricted in its expression pattern to certain neurons of the hippocampus. The hippocampus has a critical role in forming new declarative memories, which includes memories for general facts as well as knowledge of personal experiences. The importance of the hippocampus in human memory encoding is best depicted by the case of patient H.M. (recently identified as Henry Molaison) who was unable to form new long-term declarative memories after his temporal lobes (including hippocampi) were surgically removed, despite intact motor learning and intellectual abilities.[47] Subsequent lesion studies in animal models experimentally demonstrated that intact hippocampal function is required for long-term memory formation across several species. Additional aspects/types of memory are nonhippocampal dependent and are encoded by other brain regions, such as basal ganglia-dependent motor learning and amygdala-dependent auditory fear conditioning. It is now widely accepted that the hippocampus is responsible for many aspects of learning and memory including spatial, object recognition, social, and contextual memory.

Based on its high expression levels in hippocampus, we hypothesized that RGS14 likely serves a critical role in learning and memory. In studies designed to test this idea directly, we found that mice lacking RGS14 mRNA/protein (RGS14-KO) displayed enhanced performance in tests of hippocampal-dependent learning and memory compared with wild-type littermates.[7] Spatial learning was assayed using the Morris Water Maze in which the mice use visual cues to navigate the water maze and locate a hidden escape platform. Over successive trial days both groups of mice learned the task, as indicated by decreased latency to find the submerged platform, but RGS14-KO mice exhibited a marked acceleration in their rate of spatial learning. RGS14-KO mice also exhibited an enhanced ability to recognize objects they had previously encountered in the novel object test when compared with wild-type littermates. Of note, RGS14-KO mice did not differ from wild-type littermates in nonhippocampal-dependent behaviors such as open-field locomotion, startle response, and anxiety. Taken together, these findings indicate that RGS14 naturally inhibits forms of learning and memory specific to the hippocampus.

The underlying mechanism(s) by which RGS14 regulates learning and memory are unclear, although the hippocampus in particular is known for the robust capacity of its resident neurons to express plasticity—i.e., the ability of neurons to modulate their responses to neuronal activity in a synapse-specific and activity-dependent manner. Several forms of synaptic plasticity have been reported in the mammalian brain, and these phenomena are differentially regulated across postnatal development and brain regions. Long-term potentiation (LTP), the stable increase in synaptic strength in response to brief periods of synaptic stimulation, is a prevalent form of plasticity observed in the hippocampus during learning in intact animals.[48,49] Pharmacological and genetic manipulations that disrupt LTP similarly impair learning and memory formation leading to the widely accepted hypothesis that LTP is the cellular mechanism underlying memory formation and/or storage.[50] Traditional views of the hippocampus have largely focused on the trisynaptic dentate gyrus (DG)–CA3–CA1 circuit as the primary route of information flow through the hippocampus.[51] Neurons in entorhinal cortex layer II (ECII) provide excitatory input to DG granule cells, which in turn form synapses on CA3 pyramidal neurons via mossy fiber projections. These pyramidal neurons within area CA3 synapse onto CA1 pyramidal neurons via Schaffer collateral (SC) projections. High-frequency synaptic stimulation of CA3 SCs readily induces LTP in CA1 pyramidal neurons, and the cellular mechanisms behind LTP have been extensively studied at these synapses.[52] Based on the key role that LTP plays in learning and memory, we postulated that RGS14 may play a pivotal role in modulating LTP in the hippocampus.

LTP is a complex process resulting from coordinated signaling events in postsynaptic spines, compartments protruding from dendrites where most excitatory synapses form, in response to neural activity.[53] In CA1 neurons, high-frequency stimulation of presynaptic SC inputs results in the activation of postsynaptic AMPA and NMDA-type glutamate receptors (AMPARs and NMDARs, respectively). Calcium (Ca^{2+}) influx through NMDARs initiates various biochemical pathways that ultimately result in potentiation of the AMPA-mediated component of the excitatory postsynaptic current, i.e., LTP. Of note, at least two reported RGS14-binding partners (e.g., CaM, H-Ras) are activated by postsynaptic Ca^{2+} and have critical functions in LTP induction and associated spine morphology changes.[54–57] Postsynaptic Ca^{2+} entry is required for LTP induction at CA3–CA1 synapses,[58] and compelling evidence has demonstrated that postsynaptic Ca^{2+} entry activates these signaling events by activating the Ca^{2+}/calmodulin-dependent protein kinase II (CaMKII).[59] From there, the mechanisms are less clear, but one idea is that activated CaMKII phosphorylates postsynaptic AMPARs and, in addition, initiates the activity of downstream kinase cascades leading to their enhanced trafficking to the postsynaptic membrane. Several lines of evidence demonstrate that CaMKII activation is both sufficient and necessary for LTP induction in CA1 neurons and some types of learning. One of the major downstream targets of CaMKII is the H-Ras–ERK pathway.[55,57] Activity-dependent gene transcription, local protein synthesis at the stimulated synapse, and AMPAR exocytosis during LTP depend on H-Ras–ERK signaling.[56,60] H-Ras–ERK activity is required but, unlike CaMKII, is not sufficient for LTP induction in CA1 neurons. Additional biochemical pathways also have defined roles in LTP induction at hippocampal CA1 synapses, but we will only highlight cellular mechanisms relevant to RGS14 signaling and plasticity in hippocampal CA2 neurons.

The CA2 pyramidal neurons that express RGS14 also receive input from CA3 neurons, which form synapses onto the dendrites of CA2 neurons. However, this intervening area CA2 has been often overlooked and historically regarded as a "transition" zone between areas CA3 and CA1, despite early anatomical studies identifying CA2 as a separate subfield.[61] Very recent evidence has shown that CA2 pyramidal neurons possess distinct anatomical, physiological, and genetic properties from pyramidal neurons in neighboring areas CA3 and CA1.[62,63] In sharp contrast to CA1, SC synapses onto CA2 pyramidal neurons fail to support LTP following protocols that reliably induce LTP in CA1.[64] Given the enhanced learning phenotype observed in RGS14-KO mice, we designed experiments to determine if loss of RGS14 affected LTP induction in CA2 pyramidal neurons. Much to our surprise, we found that RGS14-KO mice display a robust and nascent LTP in CA2 following SC stimulation but had no difference in LTP induction in CA1, where enhanced plasticity is traditionally associated with hippocampus-dependent learning.[7] Consistent with previous reports that RGS14 negatively regulates ERK/MAPK activation, the nascent CA2 LTP observed in RGS14-KO mice could be blocked by a MEK/ERK inhibitor suggesting RGS14 may normally suppress ERK signaling to limit LTP in area CA2. These results demonstrate that RGS14 innately restricts activity-dependent plasticity in hippocampal area CA2, but not CA1. Taken together, these findings strongly suggest that the enhanced learning and memory observed with the loss of RGS14 is due to increased synaptic plasticity in CA2. This study was the first report to implicate hippocampal area CA2 and RGS14 in mediating spatial learning and novel object recognition memory. Although the cellular mechanisms regulating atypical plasticity in area CA2 and its contributions to hippocampal function and behavior are not well understood, very recent findings have elucidated additional mechanisms governing CA2 physiology and associated behaviors.[63]

3.2 Possible Mechanisms by Which RGS14 Suppresses Plasticity in Hippocampal Area CA2

Molecular mechanisms by which RGS14 suppresses LTP in CA2 neurons will now be considered. Our discussion will focus on signaling pathways that RGS14 engages and have functional roles in CA1 and CA2 neurons including Ca^{2+}, G protein, and H-Ras/ERK signaling pathways.

CA2 neurons are slightly less excitable than CA1 neurons, but Zhao *et al.* found that differences in the intrinsic electrophysiological properties relative to CA1 could not account for the plasticity-resistant phenotype observed in CA2 pyramidal neurons.[64] Therefore, some molecular difference between CA2 and CA1 is likely the factor limiting synaptic plasticity in CA2. One peculiarity of CA2 neurons is their extracellular matrix, which contains a higher concentration of proteins believed to restrict plasticity that are not found in CA3/CA1.[65–67] In addition, a number of genes are highly expressed in CA2 besides RGS14, including Amigo2, STEP, calbindin,

PCP4/Pep-19, TARP5, FGF5, neurotrophin-3, α-actinin 2, certain adenylyl cyclase isoforms (1,5,6), adenosine A1 receptor, vasopressin 1b receptor, and others.[63,68,69] Of these genes, several are members of Ca^{2+}-activated signaling pathways (calbindin, PCP4/Pep-19, certain adenylyl cyclase isoforms, vasopressin 1b receptor), are known to influence synaptic plasticity, and are likely to contribute to the atypical plasticity observed in area CA2.

Germane to the discussion are our findings that RGS14 directly engages H-Ras/ERK signaling.[19–21,70] As discussed, RGS14 binding of active H-Ras-GTP can inhibit ERK signaling. Within hippocampal neurons, ERK/MAPK signaling is essential for AMPA receptor trafficking to the postsynaptic density and the enlargement of spines associated with LTP.[55–57,60] Therefore, it is quite possible that RGS14 could inhibit LTP induction in CA2 neurons, at least in part, by inhibiting of H-Ras/ERK signaling. Studies are ongoing to test this idea directly.

Also potentially relevant are our recent findings that RGS14 can engage Ca^{2+} signaling pathways by binding CaM in a Ca^{2+}-dependent manner.[22] At this time, we do not yet know if or how RGS14/CaM interactions affect LTP, though this mechanism is under active investigation. However, LTP induction as well as ERK-mediated AMPAR trafficking in spines are both Ca^{2+}/CaM/CaMKII-dependent processes,[55,58–60] and as noted, several Ca^{2+}-binding proteins are highly expressed in CA2 pyramidal neurons.[69,71,72] Earlier studies were designed to investigate if differences in Ca^{2+} handling could explain the synaptic stability of CA2 synapses. Cell imaging studies of Ca^{2+} dynamics in dendritic spines revealed that CA2 pyramidal neurons display smaller elevations in intracellular Ca^{2+} concentrations relative to CA1 or CA3 pyramidal neurons due to higher endogenous buffering capacity and rates of Ca^{2+} extrusion.[73] Similar to loss of RGS14, raising intracellular Ca^{2+} transients in CA2 to levels comparable to CA3/CA1, either by elevating external Ca^{2+} concentration or inhibiting plasma membrane Ca^{2+} extrusion pumps, permits LTP induction. The synaptic potentiation due to high external Ca^{2+} was blocked by the NMDAR antagonist APV, indicating that Ca^{2+} influx through NMDAR is required, similar to canonical LTP observed in CA1. These findings indicate that CA2 neurons possess the cellular machinery to support LTP, but active Ca^{2+} regulation restricts the induction of activity-dependent plasticity in CA2. The robust Ca^{2+} extrusion rates and LTP-resistant phenotype can be explained, at least in part, by high expression levels of the CaM-binding protein Pep-19 in CA2 pyramidal neurons. Further, introducing camstatin, a functional peptide analog of Pep-19, into CA1 neurons resulted in much higher rates of Ca^{2+} extrusion similar to those observed in CA2 and blocked LTP induction following high-frequency stimulation of SCs. Simons *et al.* demonstrated that robust Ca^{2+} buffering and Ca^{2+} extrusion rates largely account for the limited expression of LTP in CA2 pyramidal neurons.[73] Therefore, it was an unexpected finding that RGS14-KO mice displayed a robust capacity for LTP in CA2 at the time because RGS14 had not yet been linked to Ca^{2+}-activated plasticity signaling. Our recent observations[22] that RGS14 binds Ca^{2+}/CaM suggest there may be a functional link.

RGS14 engages active Gαi/o-GTP and inactive Gαi1-GDP or Gαi3-GDP. At least one Gαi/o-coupled GPCR, specifically the A1 adenosine receptor (A1R), is highly expressed in CA2 neurons.[74] Recent studies show that antagonizing A1R is the primary mechanism by which caffeine enhances cognition.[75] A1R antagonists including caffeine have been shown to enhance LTP in hippocampal CA1 neurons,[76,77] and Simons *et al.* demonstrated that oral administration of caffeine to rats potentiates synaptic transmission in CA2 neurons, but not those in area CA1.[75] Brief application of caffeine or other more selective A1R antagonists directly to hippocampal slices also produced long-lasting synaptic potentiation postsynaptically in CA2, indicating that caffeine enhances synaptic efficacy in CA2 through blockade of A1Rs. Similar effects on synaptic strength were only observed in CA1 at substantially higher concentrations indicating that caffeine likely enhances cognition by inducing plasticity in area CA2. The A1R potentiation at CA2 synapses was not dependent on Ca^{2+}-activated pathways required for canonical LTP induced by high-frequency stimulation as it was unaffected by NMDAR antagonism, Ca^{2+} chelation, or inhibition of CaMKII. Rather, the enhancement of synaptic responses in CA2 by A1R antagonists is mediated by cAMP-dependent activation of PKA, consistent with relieving Gαi/o inhibitory effects on adenylyl cyclase by blockade of the linked GPCR. The RGS14:Gαi1 signaling complex can couple to at least one Gi/o-linked receptor (α_{2A}-AR) in exogenous expression systems.[21,33] Whether RGS14 engages A1Rs or other Gαi/o-linked GPCRs in CA2 neurons to modulate synaptic plasticity is unknown, but is a current topic of investigation.

CA2 neurons also express Gαq-linked vasopressin 1b receptor (Avpr1b) and oxytocin (Oxtr) receptors,[78,79] and stimulation of these receptors also modulates synaptic responses in area CA2.[80] The Avpr1b receptor is highly restricted in its expression to CA2 pyramidal neurons.[78] Application of specific Avpr1b agonists to rat and mouse brain slices induced synaptic potentiation of excitatory currents selectively in CA2, but not in CA1.[80] The "social" neuropeptides vasopressin and oxytocin both enhance social cognition.[81] Consistent with reports of Oxtr expression in areas CA2 and CA3 of the hippocampus, a specific oxytocin receptor agonist also enhanced excitatory synaptic responses in CA2 and CA3, but not in CA1.[79,80] Agonist stimulation of Gαq-linked GPCRs such as Avpr1b and Oxtr increases intracellular Ca^{2+} levels. In line with known roles for Ca^{2+} signaling in CA2 LTP (see above), Avpr1b and Oxtr agonists potentiate synaptic responses in CA2 through a Ca^{2+}-dependent mechanism similar to activity-dependent LTP as the effects of both agonists required synaptic stimulation during agonist treatment, NMDAR activation, postsynaptic Ca^{2+} entry, and CaMKII activity.[80] Unlike the A1R potentiation of LTP in CA2, these increases in synaptic strength were not mediated by PKA, suggesting that either Ca^{2+} or cAMP signaling events can modulate synaptic plasticity in CA2. Evidence indicates that RGS14 may engage one or both of these pathways.[22,41,63]

In summary, strong evidence suggests that the unique complement of genes enriched in RGS14-expressing CA2 pyramidal neurons gives rise to the atypical regulation of plasticity in this region. How RGS14 engages these signaling proteins and pathways to regulate LTP is an area of active investigation. While RGS14 has been shown to restrict the induction of activity-dependent LTP at SC inputs to hippocampal CA2 neurons, whether RGS14 also regulates A1R and/or Avpr1b/Oxtr-induced potentiation remains to be determined. RGS14 is well positioned to modulate these key signaling pathways required for CA2 plasticity through its known binding partners, and its capacity to functionally integrate certain components of these signaling pathways (Fig. 2). RGS14 could heighten A1R potentiation by its negative regulation of Gαi/o signaling at the RGS and GPR domains. Further, PKA activity underlying A1R potentiation could utilize RGS14 phosphorylation to prolong this plasticity by increasing the affinity of RGS14 for Gαi,[41] which could target RGS14 to inhibit subsequent A1R activation. Support for this idea is bolstered by our findings that RGS14 in complex with Gαi1 functionally associates with another Gαi/o-linked GPCR.[33] No published reports have demonstrated a functional link between RGS14 and Gαq implicated in Avpr1b/Oxtr-induced potentiation, but nonetheless, RGS14 may engage downstream partners such as CaM[22] or other Ca^{2+}-activated signaling events that mediate this form of plasticity in CA2. However, these models are currently speculative, and the involvement of RGS14 in these forms of CA2 plasticity remains to

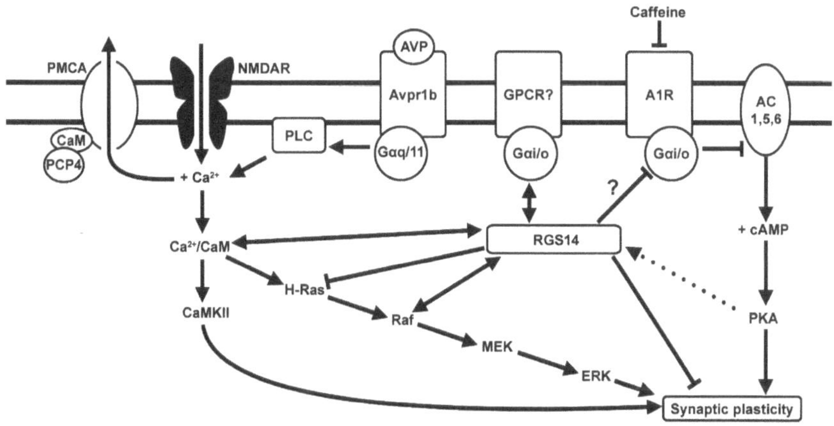

Figure 2 Postsynaptic signaling regulating plasticity in hippocampal CA2 pyramidal neurons. RGS14 is well positioned to modulate CA2 plasticity through its interactions with multiple binding partners. Avpr1b/Oxtr-potentiation signals through Gαq/11 to increase intracellular Ca^{2+} and requires synaptic stimulation and downstream Ca^{2+}/CaMKII signaling similar to canonical activity-dependent plasticity. Ca^{2+} is roughly extruded from CA2 pyramidal neurons by plasma membrane Ca^{2+} ATPases (PMCAs) that are regulated by PCP4/Pep-19 (left). RGS14, at least in part, restricts plasticity in CA2 pyramidal neurons by suppressing MAPK/ERK activity. RGS14 binds H-Ras and Raf kinases to inhibit MAPK/ERK signaling, but interactions with Gαi strongly influence this function. RGS14 also binds Ca^{2+}/CaM, and the functional consequences of this interaction on CA2 plasticity are currently under investigation (middle). Adenosine A1R antagonists potentiate synaptic responses in CA2 by relieving Gαi/o inhibition on adenylyl cyclase and downstream PKA activation, but this form of plasticity does not require Ca^{2+}-activated pathways. It is currently unknown if RGS14 modulates A1R signaling or associates with other Gi/o-linked GPCRs in area CA2. PKA phosphorylates RGS14 at two residues, but how this modification affects signaling events in hippocampal CA2 remains to be determined (right).

be determined. The features of CA2 plasticity discussed thus far have been elucidated in the context of the synaptically stable SC synapses, but recent studies using refined genetic approaches have revealed previously unknown connections to area CA2 that provide far-reaching implications for this region's function.

3.3 Connecting CA2—Redefining Anatomical Substrates of Learning

Recent studies have thrust hippocampal area CA2, the region where RGS14 is highly expressed, out of the shadows and into the spotlight.[82] Kohara *et al.* used refined cell-type-specific approaches to clarify functional circuitry linking the CA2 subfield to other brain areas.[82] These studies extended current knowledge on CA2 circuitry by identifying a new CA2-centric circuit that operates in parallel with the classic DG–CA3–CA1 trisynaptic circuit (Fig. 3). These studies revealed an alternative DG–CA2–$CA1_{deep}$ trisynaptic circuit that may encode distinct aspects of learning and memory from the traditional DG–CA3–CA1 trisynaptic circuit associated with hippocampus-dependent learning and memory.

Previous classical anatomical criteria define area CA2 as the region between subfields CA3 and CA1 containing large pyramidal cells similar in size and dendritic branching patterns to CA3, but lacking thorny excrescences indicative of innervation from the DG.[61] The CA2 subfield is also anatomically distinguished by the selective targeting of CA2 pyramidal neurons by strong projections from the supramammillary nucleus of the hypothalamus (SUM).[83] As previously discussed, CA3 pyramidal neurons send SC projections through the stratum radiatum (sr) to form synapses onto proximal dendrites of CA2 and CA1 neurons, and CA2 neurons

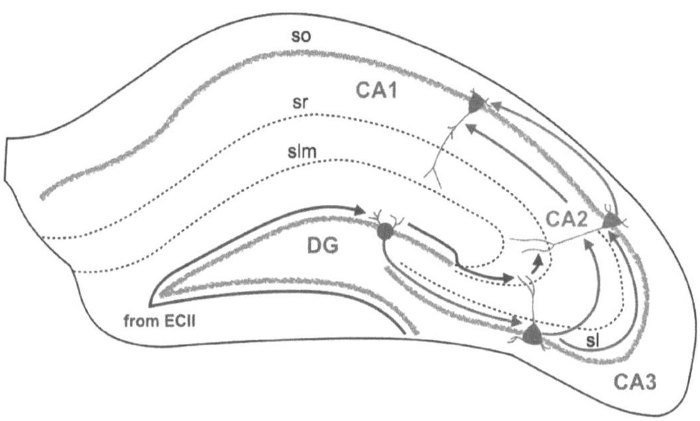

Figure 3 Area CA2 at the intersection of multiple hippocampal circuits. CA2 pyramidal neurons receive inputs from different brain regions along the proximal–distal axis of its dendrites (green). Hippocampal inputs from entorhinal cortex layer II (ECII, red) form synapses onto granule cells in the dentate gyrus (DG) and distal dendrites of CA3 and CA2 pyramidal neurons in the stratum lacunosum moleculare (slm). DG granule cells send mossy fiber terminals through the stratum lucidum (sl) that synapse onto the proximal dendrites of CA3 and CA2 pyramidal neurons (purple). CA3 pyramidal neurons form Schaffer collateral synapses (orange) onto CA2, which are usually LTP resistant, as well as CA1 in the intermediate region of the dendrite in the stratum radiatum (sr). CA2 neurons form strong monosynaptic connections onto CA1 pyramidal neurons in the stratum oriens (so). (See the color plate.)

dramatically differ from CA1 by the absence of activity-dependent plasticity at these synapses. In contrast to the synaptic stability of SC–CA2 synapses, the synapses at the distal dendrites of CA2 neurons formed by afferents from entorhinal cortex layers II and III (ECII and ECIII, respectively) projecting through the stratum lacunosum moleculare (slm) form strong monosynaptic connections that are highly plastic and support LTP induction.[84] More recent studies[82,85] have confirmed a functional direct connection from ECII to CA2 but do not support the previous finding that ECIII projects to CA2. Of note, RGS14 appears to be expressed throughout the dendritic tree, and how it restricts LTP induction at the SC synapses but not the distal entorhinal cortex synapses is unclear. Differential distribution of signaling proteins along the proximal–distal axis of CA2 dendrites or differences in presynaptic signaling may underlie these differences in plasticity.

CA2 connectivity with other brain regions has also been studied to better understand its role in hippocampal circuitry and provide insight into potential behaviors mediated by CA2. Within the hippocampus, RGS14-expressing CA2 neurons receive bilateral innervation from pyramidal neurons in ipsilateral and contralateral CA3 as well as in contralateral CA2.[85] CA2 pyramidal neurons in turn send prominent projections to neurons in all CA fields (CA3, CA2, and CA1). Thus, CA2 forms reciprocal connections with CA3 while providing dense unilateral input into CA1, providing an anatomical substrate for unidirectional information flow through the hippocampus.

CA2 is also anatomically linked with brain regions outside the hippocampus. Several groups have confirmed the SUM afferent projections to CA2, satisfying one of the classical anatomical criteria for area CA2.[74,83,85–88] A recent study identified a novel efferent projection from hippocampal CA2 to the SUM in the adult mouse brain, suggesting these two regions may reciprocally innervate each other to form a feedback loop.[85] In addition to these inputs, independent studies have found reciprocal connections between area CA2 and the septal nuclei as well as the diagonal band of Broca. Neurons in area CA2 are also innervated by the median raphe nucleus and medial and lateral ECII. The paraventricular nucleus of the hypothalamus (PVN) was recently found to provide vasopressinergic input to CA2, demonstrating a likely source of the vasopressin to activate Avpr1b in CA2.[85]

More recent studies employing genetic-based approaches have refined our understanding of CA2 circuitry. Using a modified rabies virus that could only be expressed in MECII neurons to label monosynaptic inputs,

Rowland *et al.* demonstrated a projection from CA2 pyramidal neurons indicating yet another reciprocal extrahippocampal connection.[89] Other recent studies using region-specific molecular markers found that CA2 pyramidal neurons surprisingly receive input from DG granule cells onto proximal dendrites in the stratum lucidum, despite lacking complex spines that are associated with mossy fiber synapses.[82] This study demonstrated that in the mouse, DG granule cells directly synapse onto pyramidal neurons and interneurons in CA2, and that stimulation of these mossy fiber terminals results in a functional monosynaptic connection that is dominated by feed-forward inhibition similar to SC input.[82,84] A subsequent study[90] further demonstrated that CA2 pyramidal neurons receive input from mature and immature granule cells. These immature granule cells are produced by neurogenesis in the DG, which occurs throughout life and is intimately associated with hippocampus-dependent learning and memory. The mossy fiber terminals originating from newborn DG granule cells are smaller and take longer to develop in the CA2 subfield compared to CA3. While inflammation negatively regulates immature DG cell connections with CA2 and CA3, DG–CA2 projections arising from newborn neurons can be dramatically upregulated by exercise to a greater extent than those to neighboring CA3. Thus, environmental and developmental factors, in addition to the unique molecular composition, likely play a large role in shaping CA2 circuitry.

RGS14-expressing CA2 neurons also send dense projections through the stratum oriens to preferentially target CA1 pyramidal neurons located in the deep sublayer ($CA1_{deep}$), whereas CA3 innervates CA1 pyramidal neurons in the deep and superficial sublayers equally.[82] The preferential targeting of $CA1_{deep}$ by CA2 makes sense of previous work where very few synaptically connected pairs of CA2–CA1 neurons could be identified.[84] These findings were remarkable because they highlight a *new* trisynaptic circuit wherein ECII neurons project to DG, which send mossy fiber projections to CA2, which in turn synapse onto CA1 pyramidal neurons in the deep sublayer (Fig. 3). The existence of several distinct circuits within the hippocampus might allow for separation of different input activity patterns to encode complementary or distinct aspects of memory within each circuit. Moreover, the complex corticohippocampal circuitry demonstrated in these studies allows several possible pathways for information flow through the hippocampus. For example, information could be routed through a disynaptic corticohippocampal circuit in which ECII directly stimulates CA2 that subsequently excites deep sublayer CA1 and/or reciprocally drives ECII. These reports consistently indicate that CA2 is positioned to serve as a central hub in the entorhinal–hippocampal network, and regulation of plasticity by RGS14 likely serves an important function in this network. For example, RGS14 in this context could prevent saturation/overlearning or preserve signal fidelity in this circuit.

The anatomical links described between CA2 and other regions provide clues into potential function of area CA2 and RGS14 in mediating animal behavior. Reciprocal connectivity between CA2 and cholinergic projections (septal nuclei and diagonal band) may play an important role in attention and memory. Input from PVN demonstrates CA2 may integrate social information with other contextual information routed through the hippocampal circuit. Thus, CA2 is poised to regulate theta rhythm in the hippocampus observed during exploratory behavior through connections with SUM. Consistent with our finding that robust CA2 plasticity in RGS14-KO mice correlates with enhanced spatial learning, CA2 connectivity to ECII could modulate grid cells, which orient an animal's location in space. Stellate cells within MECII are putative grid cells that are responsible for forming a mental framework to navigate an environment[91–93]; elimination of hippocampal excitatory input to grid cells abolishes the ability of these cells to form a grid pattern. RGS14-expressing neurons within area CA2 are the obvious candidates for providing excitatory drive to grid cells as the vast majority of hippocampal inputs to MECII arise from CA2.[89] The specific contribution of CA2 pyramidal neurons and the expression of RGS14 therein to grid cell function will provide great insight into the neural mechanisms underlying spatial learning and memory.

In summary, the relative stability at CA3–CA2 synapses and the pivotal role of RGS14 in this regard may provide a mechanism for tightly regulating plasticity at those synapses that can only be induced under specific circumstances.[63] We envision a circuit whereby CA3 inputs to CA2 may modulate rather than drive responses from CA2, thereby acting like a filter to preserve fidelity in the circuit and/or acting to differentially route information through the putative circuits within the hippocampus. Plasticity in CA2 may only occur when convergent input (e.g., vasopressin release from PVN during social interaction) allows selective encoding of some types of memory. RGS14 plays a central role in restricting plasticity at SC synapses onto CA2 neurons and may therefore serve as a molecular regulator of memory storage at these specific synapses. Future studies are necessary to determine if RGS14 similarly regulates CA2 plasticity at its other synaptic inputs and modulates additional forms of learning and memory linked to CA2.

3.4 Do RGS14-Expressing CA2 Neurons Control Social Behavior?

Hippocampal CA2 has only recently been a focus of study; yet compelling evidence supports a critical role for area CA2 in social behavior. In rodents, hippocampal CA2 neurons highly express two GPCRs closely linked to social behavior, vasopressin 1b receptors (Avpr1b) and the oxytocin receptor (Oxtr).[78,79] Similar to loss of RGS14,[7] stimulation of Avpr1b induces long-lasting potentiation specifically in CA2 pyramidal neurons.[80] Given that the Avpr1b is largely restricted in its expression to CA2 pyramidal neurons, behavioral studies of mice lacking Avpr1b (Avpr1b KO) provide valuable insight into the function of CA2 in other behaviors. Avpr1b KO mice display normal sensorimotor function, olfactory discrimination, exploratory behavior, and spatial memory. However, Avpr1b KO mice are deficient in measures of sociability and social novelty,[94] showing a lack of interest in social interaction or motivation. In tests of social novelty, Avpr1b KO mice fail to distinguish between novel and familiar mice, indicating they either do not prefer the novel animal or do not remember the familiar conspecific. These studies demonstrate that Avpr1b is required for normal social interactions and memory, consistent with the role of vasopressin as a social neuropeptide.

DeVito *et al.* investigated whether hippocampus-dependent contextual learning and memory were perturbed in Avpr1b KO mice.[94] These experiments revealed that Avpr1b KO mice had a specific impairment in the temporal order of events. In the "what–where–when" memory task, Avpr1b KO mice exhibited normal performance for where in the arena they explored the objects, confirming previous studies demonstrating intact spatial memory in these animals. However, Avpr1b KO mice were impaired in recognizing previously encountered objects and failed to remember the temporal order in which the objects were presented relative to wild-type littermates. In the object–trace–odor assay, Avpr1b KO mice could associate odors with objects and demonstrated intact relational memory, but they failed to discriminate an odor associated with a previously encountered object over a temporal delay. Taken together, these studies suggest that the Avpr1b, likely in CA2, is required for the proper temporal association of episodic events.

In addition to social recognition, the appropriate expression of aggression is another key component of mammalian social behavior. Male and female Avpr1b KO mice display reduced territorial and maternal aggression, respectively, and specific pharmacological antagonism of the Avpr1b mirrors this behavior.[95] Reintroduction of Avpr1b by lentiviral injection into CA2 pyramidal neurons of Avpr1b KO mice partially rescued Avpr1b expression and restored male aggression in response to intruder attacks.[80] Additional control behavioral studies proved this effect was specific to aggression and not, for example, anxiety-like behavior. These findings demonstrate that expression of Avpr1b in CA2 underlies the proper expression of aggressive behaviors in rodents. Additional behavioral analyses will be required to discern other facets of behavior mediated by Avpr1b in CA2 and if RGS14 contributes to these behaviors.

In further support of a key role for CA2 in social behaviors, a recent study found that CA2 is necessary for social memory.[96] Silencing of CA2 pyramidal neuron output by expression of tetanus neurotoxin resulted in profound deficits in social memory. Although sociability remained intact, silencing CA2 resulted in failure to display a preference for novel mouse over a familiar littermate indicating lack of social novelty and further that CA2 activity is required to encode social memory. The effects of CA2 silencing were specific to social recognition/memory as there were no differences in a battery of other hippocampal and nonhippocampal behaviors. Among these behaviors not affected by CA2 silencing were novel object recognition and spatial learning and memory assayed by Morris Water Maze, despite a trend for these mice to learn the task slower. Because loss of RGS14 in CA2 neurons results in robust plasticity correlating with enhanced performance in these tasks,[7] we hypothesized that area CA2 functions in spatial learning and that silencing CA2 synaptic transmission would conversely cause impairments in spatial learning and object recognition. Thus, it is possible that spatial information is routed through alternative circuits when CA2 is silenced. Alternatively, the lack of effect on spatial learning could be due to an incomplete silencing of CA2 over the entire extent of the hippocampus.

Taken together, these initial findings suggest that RGS14-expressing neurons of hippocampal area CA2 encode spatial,[7] temporal,[94] and social information.[80,94,96] Future studies are necessary to determine the precise mnemonic functions of hippocampal CA2 and assess the contribution of RGS14 (and other genes enriched in CA2) to these processes. The distinct molecular makeup of CA2 pyramidal neurons provides promising avenues to study CA2 function using genetic-based targeting approaches. Selective deletion of genes in CA2 and manipulation of CA2 activity *in vivo*, e.g., employing optogenetic stimulation or DREADDs, as well as *in vivo* recordings during behavior, will offer great insight into the type(s) of information processed by this region and its function in learning and memory. Understanding the normal functions of CA2 will also provide clues to the contribution of CA2 dysfunction in human neurological disorders.

4. POTENTIAL ROLES FOR RGS14 AND CA2 IN HUMAN BEHAVIOR AND DISEASE

In humans and nonhuman primates, RGS14 is expressed abundantly in CA2 hippocampal neurons much like rodents,[97,98] and hippocampal dysfunction characterized by cognitive impairments is a central feature of numerous human neuropsychiatric diseases. Early evidence demonstrating that area CA2 is a distinct hippocampal subfield, rather than an intermingling of CA3 and CA1, originated from reports that CA2 neurons display unique pathology in some human neurological conditions and are resistant to neuronal injury in others.[63,99–102] In the normal aged human and Alzheimer's disease (AD) patient brain, area CA2 is resistant to tau neurofibrillary tangle formation, a hallmark of AD.[103] Conversely, neurofibrillary pathology is selectively observed in the CA2 subfield in rare cases of specific human tauopathies. Within the hippocampus, autopsy tissue from human schizophrenic patients features a prominent and selective loss of parvalbumin immunoreactivity[101] (a molecular marker for a subtype of inhibitory interneurons), decreased size of pyramidal cells,[104] and reduced AMPAR binding in the CA2 subregion.[105] How these CA2-specific alterations contribute to the symptoms of schizophrenia is currently unknown, but further study of the mnemonic functions of CA2 could provide great insight into its role in the disease etiology.

Hippocampal CA2 also displays resilience in response to neuronal insults. Unlike surrounding hippocampal subfields, CA2 survives in several models of hypoxia/ischemia,[99,100] and CA2 remarkably differs from CA1/3 in that no detectable neuronal loss is observed in human subjects following blunt head injury.[106] Perhaps the strongest evidence of CA2's resilience comes from studies demonstrating the diminished susceptibility to seizure-induced cell loss in human epileptic patients and experimental models of temporal lobe epilepsy.[102,107]

Consistent with its resistance to damage to hypoxia and ischemia, CA2 neurons also are resistant to cell loss following epileptic seizures,[102,107,110–112] and several signaling proteins enriched in hippocampal CA2, possibly including RGS14, may act in concert to confer protection. The high expression levels of Ca^{2+} buffering proteins and active

Ca^{2+} handling processes in CA2 pyramidal neurons are believed to reduce the seizure-induced activation of Ca^{2+}-dependent apoptotic pathways resulting in limited cell death.[113] RGS14 may contribute to the unique Ca^{2+} buffering and extrusion properties in CA2 neurons by its binding to Ca^{2+}/CaM, similar to PCP4/Pep-19. Adenosine is known to have anticonvulsant effects, and the preferential sparing of CA2 in models of temporal lobe epilepsy has also been attributed to activation of A1R by adenosine released during seizures.[74,114] RGS14 may contribute to adenosine's actions in CA2 by engaging Gαi signaling pathways downstream of A1R. Amigo2, a cell surface adhesion protein that also is highly enriched in and restricted in its expression to area CA2,[115] may contribute to the neuroprotection of pyramidal neurons similar to its capacity to promote neuronal survival in response to Ca^{2+} signals in the cerebellum.[116] Moreover, decreased levels of ERK phosphorylation observed in CA2 compared to CA1 after experimentally induced seizure may be a critical component of CA2 survival strategy. RGS14 binding of active H-Ras to suppress ERK signaling may contribute to the reduced ERK activation in this case. How RGS14 engages these proteins and pathways to contribute to CA2 synaptic stability and survival is currently unknown, but ongoing studies are assessing the functional role of RGS14 in seizure activity and cell survival.

What properties bestow this robust neuroprotective phenotype and what is the significance of selectively sparing CA2? As discussed above, the unique molecular composition of area CA2 and its resistance to LTP induction, due to RGS14 and robust calcium handling properties, are proposed to underlie the distinct pathology and resistance to cell death. Aberrant hyperphosphorylation of tau protein is a central mechanism in neurofibrillary tangle degeneration, and the lack of tau pathology in CA2 may be attributed to the enrichment of phosphatases such as STEP (striatal-enriched protein tyrosine phosphatase), which might impede tau hyperphosphorylation. The enrichment of the well-known neuroprotective agent neurotrophin-3 in CA2 may also confer resistance to damage in ischemic events as it has been found to protect cultured hippocampal neurons from these insults.[108,109]

A deep literature on neuronal plasticity shows that brain regions exhibiting synaptic plasticity do so at the expense of increased susceptibility to damage, indicating a trade-off between these two features. Consistent with this idea, the complement of signaling proteins that uniquely limit plasticity in area CA2, most notably RGS14, may serve an adaptive mechanism to ensure this region's survival from neuronal injury. This hypothesis would suggest that although CA2 may store memories under specific circumstances, the information encoded within this region serves a critical function that mandates its survival. We expect that area CA2 will be implicated in additional human neuropsychiatric conditions as mnemonic functions of this region are further defined. The recently described role of area CA2 in social memory[80,94,96,117] suggests that dysfunction in CA2 could be a central feature in human disorders characterized by impaired social information processing, such as autism spectrum disorders, Rett syndrome, and schizophrenia.

Based on these findings, we hypothesize that RGS14 restricts CA2 plasticity to allow selective learning and memory encoding under specific conditions, and that its presence is required to discriminate which forms of learning and memory to encode. Of note, microarray data confirms that RGS14 is highly expressed in nonhuman primate and human CA2,[97,98] and this conserved expression pattern could suggest that RGS14 also suppresses LTP in primate CA2. If RGS14 indeed plays a pivotal role in selective memory storage, then genetic mutations that eliminate RGS14 function could either enhance cognitive function or, alternatively, be maladaptive by potentially allowing indiscriminant memory storage or cognitive inflexibility. Consistent with this notion, RGS14 was recently identified as a candidate gene involved in fear learning, suggesting it could have a potential role in human posttraumatic stress disorder.[118] Studies are in progress to define a functional role for RGS14 in these behaviors and associated disorders.

4.1 RGS14 Expression During Early Postnatal Development: The Aging Conspiracy Against Plasticity in CA2

Heightened synaptic plasticity is prevalent during critical periods of postnatal brain development when sensory experience influences the maturation of neural circuitry. Early life events have profound impacts on the hippocampus that can persist into adulthood, and hippocampal-dependent processing of spatial and social information is required for adaptation and survival in early development. Given that RGS14 suppresses hippocampus-dependent learning and memory, we investigated if RGS14 expression is regulated during postnatal mouse brain development.[43] We found that RGS14 protein is undetectable at birth (P0) with very low levels of mRNA present in brain. RGS14 mRNA and protein expression levels are dramatically upregulated in brain during postnatal mouse brain development, with protein first detected at P7, and both increasing over time until reaching highest sustained levels in adulthood. This expression pattern was mirrored by upregulated immunoreactivity for RGS14 protein observed in hippocampal

CA2. Taken together with behaviors recently linked to area CA2, these data suggest that the absence of RGS14 protein may permit enhanced hippocampus-dependent learning during the first weeks of life, such as maternal bonding and acquiring spatial memory of the environment. Subsequent upregulated expression of RGS14 beginning at P7 could serve to selectively filter episodic learning and memory storage driven by experience.

The presence of RGS14 expression coincides with the most active period of synapse formation and pruning,[119] suggesting that the appearance of RGS14 and other CA2-enriched signaling proteins could possibly shape the maturation of hippocampal CA2 circuitry. The CaM-binding protein PCP4/Pep-19 is essential for synaptic plasticity in area CA2,[73] and its developmental expression pattern in CA2 is strikingly similar to that of RGS14.[120] The enrichment of PCP4 in area CA2 underlies, at least in part, the robust Ca^{2+} buffering and extrusion properties.[73] The development expression pattern of A1R in rodent hippocampal CA2 also resembles the trends for RGS14 and PCP4.[74] Further, perineuronal nets surrounding CA2 neurons have been shown to increase during development and restrict plasticity in other brain regions.[121] The coincident appearance and the expression patterns of these genes in area CA2 suggest that plasticity is developmentally regulated in this region to influence hippocampal circuitry and learning. Another possible interpretation of the concerted expression of these proteins is that they serve a neuroprotective function to stabilize and preserve hippocampal CA2 throughout life. While these findings offer developmental insight into CA2, these studies also revealed RGS14 expression in brain regions outside of hippocampal CA2.

4.2 RGS14: More than Just a Suppressor of Learning and Memory?

While RGS14 expression is highest in hippocampal CA2, RGS14 protein is also expressed in other regions of the mouse brain, providing insight to its other potential functions in brain.[43] In the hippocampus, RGS14 expression increases during postnatal development in fasciola cinerea (FC)—the region located at the midline of the brain neighboring CA1. While the specific functions of the FC are currently unknown, the molecular markers used to identify hippocampal CA2 are curiously also expressed in the FC. A recent study suggests that area CA2 and FC comprise one region in the anterior hippocampus, which becomes separated in the posterior segment of the hippocampus into the medial FC and lateral CA2 by the interjection of CA1.[115] Sporadic immunolabeling of RGS14 also is observed in CA1 neurons, although its expression is not consistently detected in this region. RGS14 functions in regions of adult brain outside area CA2 currently are unknown, but its expression pattern suggests additional functions are likely.

Outside the hippocampus, RGS14 expression is also upregulated during postnatal development in pyramidal neurons of the anterior olfactory nucleus and piriform cortex. These brain regions are included in a collective group of brain structures referred to as the primary olfactory cortex. In mammals, the olfactory bulb initially processes odorants and sends direct projections to the primary olfactory cortex. The primary olfactory cortex processes input provided by the olfactory bulb and associates odorants with episodic events.[122] RGS14 protein is found in neurons with pyramidal morphology in the orbital and entorhinal cortices, both of which receive input from primary olfactory cortex. Thus, RGS14 is well positioned to modulate olfactory processing in mice and the association of these stimuli with specific memories. Olfaction is an essential component of mammalian social behavior, and ongoing experiments are assessing the role of RGS14 in social learning and odorant association.

In contrast to our findings that RGS14 naturally inhibits novel object recognition,[7] one study reported that RGS14 enhances object recognition memory when introduced as recombinant protein into other brain areas.[123] RGS14 is not natively expressed in rodent visual cortex;[43] however, López-Aranda *et al.* reported that ectopic overexpression of RGS14 in this region in rats promotes the conversion of short-term object recognition memory to stable, long-term memory. This puzzling finding that recombinant RGS14 enhances recognition memory differs from our work demonstrating that native RGS14 inhibits novel objection recognition memory by restricting LTP in CA2 neurons.[7] It is possible that unnatural expression of RGS14 modulates this form of memory by perturbing G protein, H-Ras/MAPK, and/or Ca^{2+} signaling leading to augmentation of plasticity in visual cortex neurons. How RGS14 alters the activity of neurons in visual cortex and the cellular mechanisms underlying this effect are currently unknown.

Despite these findings with ectopically expressed RGS14, the question still remains whether native RGS14 suppresses activity-dependent plasticity (i.e., LTP) in neuronal populations outside of hippocampal CA2. One striking observation is that RGS14 protein subcellular localization differs dramatically between CA2 pyramidal neurons and other regions, despite consistent labeling of neurons with pyramidal morphology.[43] In CA2 pyramidal neurons, RGS14 immunoreactivity intensely labels the postsynaptic dendrites and spines; by contrast, RGS14 protein is localized to the soma and apical dendrites in pyramidal neurons in other brain regions during development.[43] The lower levels of RGS14 immunoreactivity and differences in subcellular localization suggest RGS14 may serve distinct functions in areas outside of CA2 during postnatal development.

5. SUMMARY AND PERSPECTIVES

The findings discussed here highlight the defined roles of RGS14 in cell signaling, hippocampus physiology, and animal behavior. As with other members of the RGS protein family, RGS14 negatively regulates canonical GPCR/G protein signaling by serving as a GAP for Gα-GTP subunits through its conserved RGS domain. The unusual domain structure of RGS14 allows it to bridge these conventional G protein pathways with unconventional G protein signaling through its GPR motif as well as MAPK signaling by virtue of its tandem RBDs. RGS14 is able to functionally integrate these pathways in cells to impact specific downstream signaling events, and RGS14 function is subject to complex regulation by binding partners and other factors. We reviewed the defined physiological role of RGS14 as an inherent factor restricting plasticity in the peculiar CA2 pyramidal neurons within the hippocampus. These recently distinguished neurons were historically thought to be a mere transition zone in the hippocampus, despite distinct molecular composition and differential pathology in neurological conditions. However, mice lacking RGS14 display a nascent capacity for plasticity specific to CA2 and also have enhanced spatial learning and object recognition memory. Very recent reports have linked hippocampal CA2 to additional behaviors including temporal order of events and social behavior. Future behavioral studies will identify potential roles for RGS14 in these behaviors mediated by CA2 plasticity. Further elucidating the functional significance of RGS14 expression in other brain areas and peripheral tissues could yield even greater insight into this unusual RGS protein.

Still the question remains as to why a gene would exist that seemingly only functions to inhibit learning and memory. Similar to rodents, RGS14 expression levels are highest in human and nonhuman primate CA2,[97,98] suggesting it may serve an evolutionarily conserved function in this hippocampal subregion. We propose that RGS14 acts as a tightly regulated filter to selectively permit memory storage in CA2 under specific conditions. Loss of RGS14 function in CA2 of humans could potentially result in enhanced cognition or, alternatively, maladaptive consequences such as indiscriminant memory encoding or runaway excitation of CA2 neurons. Moreover, RGS14 and/or CA2 dysfunction could have consequences on social information processing. Consistent with this idea, variations in the copy number of alleles for the chromosomal location containing the human *RGS14* gene are clinically linked to autism spectrum disorder as well as developmental delay.[124] These findings underscore the importance of RGS14 and its host CA2 neurons in human cognitive function.

ACKNOWLEDGMENTS

P.R.E. was supported by a National Research Service Award predoctoral fellowship from the National Institutes of Health/National Institute of Neurological Disorders and Stroke (1F31 NS086174). J.R.H. was supported by grants from the National Institutes of Health/National Institute of Neurological Disorders and Stroke (5R01 NS37112 and 1R21NS074975). S.M.D. was supported by the Intramural Research Program of the National Institute of Environmental Health Sciences, National Institutes of Health (Z01-ES-100221).

REFERENCES

1. De Vries L, Zheng B, Fischer T, Elenko E, Farquhar MG. The regulator of G protein signaling family. *Annu Rev Pharmacol Toxicol*. 2000;40:235–271. http://dx.doi.org/10.1146/annurev.pharmtox.40.1.235.
2. Hollinger S, Hepler JR. Cellular regulation of RGS proteins: modulators and integrators of G protein signaling. *Pharmacol Rev*. 2002;54(3):527–559. Available at: http://www.ncbi.nlm.nih.gov/pubmed/12223533, Accessed March 26, 2013.
3. Ross EM, Wilkie TM. GTPase-activating proteins for heterotrimeric G proteins: regulators of G protein signaling (RGS) and RGS-like proteins. *Annu Rev Biochem*. 2000;69:795–827. http://dx.doi.org/10.1146/annurev.biochem.69.1.795.
4. Willars GB. Mammalian RGS, proteins: multifunctional regulators of cellular signalling. *Semin Cell Dev Biol*. 2006;17(3):363–376. http://dx.doi.org/10.1016/j.semcdb.2006.03.005.
5. Abramow-Newerly M, Roy AA, Nunn C, Chidiac P. RGS proteins have a signalling complex: interactions between RGS proteins and GPCRs, effectors, and auxiliary proteins. *Cell Signal*. 2006;18(5):579–591. http://dx.doi.org/10.1016/j.cellsig.2005.08.010.
6. Vellano CP, Lee SE, Dudek SM, Hepler JR. RGS14 at the interface of hippocampal signaling and synaptic plasticity. *Trends Pharmacol Sci*. 2011;32(11):666–674. http://dx.doi.org/10.1016/j.tips.2011.07.005.
7. Lee SE, Simons SB, Heldt SA, et al. RGS14 is a natural suppressor of both synaptic plasticity in CA2 neurons and hippocampal-based learning and memory. *Proc Natl Acad Sci USA*. 2010;107(39):16994–16998. http://dx.doi.org/10.1073/pnas.1005362107.
8. Hepler JR, Gilman AG. G proteins. *Trends Biochem Sci*. 1992;17(10):383–387. Available at: http://www.ncbi.nlm.nih.gov/pubmed/1455506, Accessed December 17, 2014.
9. Simon MI, Strathmann MP, Gautam N. Diversity of G proteins in signal transduction. *Science*. 1991;252(5007):802–808. Available at: http://www.ncbi.nlm.nih.gov/pubmed/1902986, Accessed December 8, 2014.

10. Hamm HE, Gilchrist A. Heterotrimeric G proteins. *Curr Opin Cell Biol.* 1996;8(2):189–196. Available at: http://www.ncbi.nlm.nih.gov/pubmed/8791425, Accessed December 18, 2014.
11. Wettschureck N, Offermanns S. Mammalian G proteins and their cell type specific functions. *Physiol Rev.* 2005;85(4):1159–1204. http://dx.doi.org/10.1152/physrev.00003.2005.
12. Hepler JR. Emerging roles for RGS proteins in cell signalling. *Trends Pharmacol Sci.* 1999;20(9):376–382. Available at: http://www.ncbi.nlm.nih.gov/pubmed/10462761, Accessed December 17, 2014.
13. Neitzel KL, Hepler JR. Cellular mechanisms that determine selective RGS protein regulation of G protein-coupled receptor signaling. *Semin Cell Dev Biol.* 2006;17(3):383–389. http://dx.doi.org/10.1016/j.semcdb.2006.03.002.
14. McCoy KL, Hepler JR. Regulators of G protein signaling proteins as central components of G protein-coupled receptor signaling complexes. *Prog Mol Biol Transl Sci.* 2009;86:49–74. http://dx.doi.org/10.1016/S1877-1173(09)86003-1.
15. Snow BE, Antonio L, Suggs S, Gutstein HB, Siderovski DP. Molecular cloning and expression analysis of rat Rgs12 and Rgs14. *Biochem Biophys Res Commun.* 1997;233(3):770–777. http://dx.doi.org/10.1006/bbrc.1997.6537.
16. Traver S, Bidot C, Spassky N, et al. RGS14 is a novel Rap effector that preferentially regulates the GTPase activity of galphao. *Biochem J.* 2000;350(pt 1): 19–29. Available at: http://www.pubmedcentral.nih.gov/articlerender.fcgi?artid=1221220&tool=pmcentrez&rendertype=abstract, Accessed March 21, 2013.
17. Hollinger S, Taylor JB, Goldman EH, Hepler JR. RGS14 is a bifunctional regulator of Galphai/o activity that exists in multiple populations in brain. *J Neurochem.* 2001;79(5):941–949. Available at: http://www.ncbi.nlm.nih.gov/pubmed/11739605, Accessed March 21, 2013.
18. Cho H, Kozasa T, Takekoshi K, De Gunzburg J, Kehrl JH. RGS14, a GTPase-activating protein for Gialpha, attenuates Gialpha- and G13alpha-mediated signaling pathways. *Mol Pharmacol.* 2000;58(3):569–576. Available at: http://www.ncbi.nlm.nih.gov/pubmed/10953050, Accessed March 26, 2013.
19. Kiel C, Wohlgemuth S, Rousseau F, et al. Recognizing and defining true Ras binding domains II: in silico prediction based on homology modelling and energy calculations. *J Mol Biol.* 2005;348(3):759–775. http://dx.doi.org/10.1016/j.jmb.2005.02.046.
20. Shu F, Ramineni S, Hepler JR. RGS14 is a multifunctional scaffold that integrates G protein and Ras/Raf MAPkinase signalling pathways. *Cell Signal.* 2010; 22(3):366–376. http://dx.doi.org/10.1016/j.cellsig.2009.10.005.
21. Vellano CP, Brown NE, Blumer JB, Hepler JR. Assembly and function of the regulator of G protein signaling 14 (RGS14)·H-Ras signaling complex in live cells are regulated by Gαi1 and Gαi-linked G protein-coupled receptors. *J Biol Chem.* 2013; 288(5):3620–3631. http://dx.doi.org/10.1074/jbc.M112.440057.
22. Evans PR, Hepler JR. Regulator of G protein signaling 14 (RGS14) interacts with calmodulin (CaM) in a calcium-dependent manner. In: *Society for Neuroscience*. New Orleans, LA; 2012:44.04.
23. Kimple RJ, De Vries L, Tronchère H, et al. RGS12 and RGS14 GoLoco motifs are Galpha(i) interaction sites with guanine nucleotide dissociation inhibitor activity. *J Biol Chem.* 2001;276(31):29275–29281. http://dx.doi.org/10.1074/jbc.M103208200.
24. Mittal V, Linder ME. The RGS14 GoLoco domain discriminates among Galphai isoforms. *J Biol Chem.* 2004;279(45):46772–46778. http://dx.doi.org/10.1074/jbc.M407409200.
25. Shu F, Ramineni S, Amyot W, Hepler JR. Selective interactions between Gialpha1 and Gialpha3 and the GoLoco/GPR domain of RGS14 influence its dynamic subcellular

localization. *Cell Signal*. 2007;19(1):163–176. http://dx.doi.org/10.1016/j.cellsig.2006.06.002.
26. Vellano CP, Shu F-J, Ramineni S, Yates CK, Tall GG, Hepler JR. Activation of the regulator of G protein signaling 14-Gαi1-GDP signaling complex is regulated by resistance to inhibitors of cholinesterase-8A. *Biochemistry*. 2011;50(5):752–762. http://dx.doi.org/10.1021/bi101910n.
27. Zhao P, Nunn C, Ramineni S, Hepler JR, Chidiac P. The Ras-binding domain region of RGS14 regulates its functional interactions with heterotrimeric G proteins. *J Cell Biochem*. 2013;114(6):1414–1423. http://dx.doi.org/10.1002/jcb.24483.
28. Hepler JR, Cladman W, Ramineni S, Hollinger S, Chidiac P. Novel activity of RGS14 on Goalpha and Gialpha nucleotide binding and hydrolysis distinct from its RGS domain and GDI activity. *Biochemistry*. 2005;44(14):5495–5502. http://dx.doi.org/10.1021/bi048359d.
29. Blumer JB, Lanier SM. Activators of G protein signaling exhibit broad functionality and define a distinct core signaling triad. *Mol Pharmacol*. 2014;85(3):388–396. http://dx.doi.org/10.1124/mol.113.090068.
30. Sato M, Blumer JB, Simon V, Lanier SM. Accessory proteins for G proteins: partners in signaling. *Annu Rev Pharmacol Toxicol*. 2006;46:151–187. http://dx.doi.org/10.1146/annurev.pharmtox.46.120604.141115.
31. Willard FS, Kimple RJ, Siderovski DP. Return of the GDI: the GoLoco motif in cell division. *Annu Rev Biochem*. 2004;73:925–951. http://dx.doi.org/10.1146/annurev.biochem.73.011303.073756.
32. Hampoelz B, Knoblich JA. Heterotrimeric G proteins: new tricks for an old dog. *Cell*. 2004;119(4):453–456. http://dx.doi.org/10.1016/j.cell.2004.10.025.
33. Vellano CP, Maher EM, Hepler JR, Blumer JB. G protein-coupled receptors and resistance to inhibitors of cholinesterase-8A (Ric-8A) both regulate the regulator of g protein signaling 14 RGS14·Gαi1 complex in live cells. *J Biol Chem*. 2011;286(44):38659–38669. http://dx.doi.org/10.1074/jbc.M111.274928.
34. Cho H, Kim D-U, Kehrl JH. RGS14 is a centrosomal and nuclear cytoplasmic shuttling protein that traffics to promyelocytic leukemia nuclear bodies following heat shock. *J Biol Chem*. 2005;280(1):805–814. http://dx.doi.org/10.1074/jbc.M408163200.
35. Tall GG, Krumins AM, Gilman AG. Mammalian Ric-8A (synembryn) is a heterotrimeric Galpha protein guanine nucleotide exchange factor. *J Biol Chem*. 2003;278(10):8356–8362. http://dx.doi.org/10.1074/jbc.M211862200.
36. Tall GG, Gilman AG. Resistance to inhibitors of cholinesterase 8A catalyzes release of Galphai-GTP and nuclear mitotic apparatus protein (NuMA) from NuMA/LGN/Galphai-GDP complexes. *Proc Natl Acad Sci USA*. 2005;102(46):16584–16589. http://dx.doi.org/10.1073/pnas.0508306102.
37. Thomas CJ, Tall GG, Adhikari A, Sprang SR. Ric-8A catalyzes guanine nucleotide exchange on Galphai1 bound to the GPR/GoLoco exchange inhibitor AGS3. *J Biol Chem*. 2008;283(34):23150–23160. http://dx.doi.org/10.1074/jbc.M802422200.
38. Oner SS, An N, Vural A, et al. Regulation of the AGS3·G{alpha}i signaling complex by a seven-transmembrane span receptor. *J Biol Chem*. 2010;285(44):33949–33958. http://dx.doi.org/10.1074/jbc.M110.138073.
39. Oner SS, Maher EM, Breton B, Bouvier M, Blumer JB. Receptor-regulated interaction of activator of G-protein signaling-4 and Galphai. *J Biol Chem*. 2010;285(27):20588–20594. http://dx.doi.org/10.1074/jbc.C109.088070.
40. Hollinger S, Hepler JR. Methods for measuring RGS protein phosphorylation by G protein-regulated kinases. *Methods Mol Biol*. 2004;237:205–219. Available at: http://www.ncbi.nlm.nih.gov/pubmed/14501052, Accessed March 21, 2013.

41. Hollinger S, Ramineni S, Hepler JR. Phosphorylation of RGS14 by protein kinase A potentiates its activity toward Galphai. *Biochemistry*. 2003;42(3):811–819. http://dx.doi.org/10.1021/bi026664y.
42. Hornbeck PV, Kornhauser JM, Tkachev S, et al. PhosphoSitePlus: a comprehensive resource for investigating the structure and function of experimentally determined post-translational modifications in man and mouse. *Nucleic Acids Res*. 2012;40(Database issue):D261–D270. http://dx.doi.org/10.1093/nar/gkr1122.
43. Evans PR, Lee SE, Smith Y, Hepler JR. Postnatal developmental expression of regulator of G protein signaling 14 (RGS14) in the mouse brain. *J Comp Neurol*. 2014;522(1):186–203. http://dx.doi.org/10.1002/cne.23395.
44. Larminie C, Murdock P, Walhin J-P, et al. Selective expression of regulators of G-protein signaling (RGS) in the human central nervous system. *Brain Res Mol Brain Res*. 2004;122(1):24–34. http://dx.doi.org/10.1016/j.molbrainres.2003.11.014.
45. López-Aranda MF, Acevedo MJ, Carballo FJ, Gutiérrez A, Khan ZU. Localization of the GoLoco motif carrier regulator of G-protein signalling 12 and 14 proteins in monkey and rat brain. *Eur J Neurosci*. 2006;23(11):2971–2982. http://dx.doi.org/10.1111/j.1460-9568.2006.04838.x.
46. Grafstein-Dunn E, Young KH, Cockett MI, Khawaja XZ. Regional distribution of regulators of G-protein signaling (RGS) 1, 2, 13, 14, 16, and GAIP messenger ribonucleic acids by in situ hybridization in rat brain. *Brain Res Mol Brain Res*. 2001;88(1–2):113–123. Available at: http://www.ncbi.nlm.nih.gov/pubmed/11295237, Accessed December 2, 2014.
47. Corkin S. What's new with the amnesic patient H.M.? *Nat Rev Neurosci*. 2002;3(2):153–160. http://dx.doi.org/10.1038/nrn726.
48. Whitlock JR, Heynen AJ, Shuler MG, Bear MF. Learning induces long-term potentiation in the hippocampus. *Science*. 2006;313(5790):1093–1097. http://dx.doi.org/10.1126/science.1128134.
49. Gruart A, Muñoz MD, Delgado-García JM. Involvement of the CA3-CA1 synapse in the acquisition of associative learning in behaving mice. *J Neurosci*. 2006;26(4):1077–1087. http://dx.doi.org/10.1523/JNEUROSCI.2834-05.2006.
50. Grant SG, Silva AJ. Targeting learning. *Trends Neurosci*. 1994;17(2):71–75. Available at: http://www.ncbi.nlm.nih.gov/pubmed/7512771, Accessed December 2, 2014.
51. Van Strien NM, Cappaert NLM, Witter MP. The anatomy of memory: an interactive overview of the parahippocampal-hippocampal network. *Nat Rev Neurosci*. 2009;10(4):272–282. http://dx.doi.org/10.1038/nrn2614.
52. Kerchner GA, Nicoll RA. Silent synapses and the emergence of a postsynaptic mechanism for LTP. *Nat Rev Neurosci*. 2008;9(11):813–825. http://dx.doi.org/10.1038/nrn2501.
53. Murakoshi H, Yasuda R. Postsynaptic signaling during plasticity of dendritic spines. *Trends Neurosci*. 2012;35(2):135–143. http://dx.doi.org/10.1016/j.tins.2011.12.002.
54. Xia Z, Storm DR. The role of calmodulin as a signal integrator for synaptic plasticity. *Nat Rev Neurosci*. 2005;6(4):267–276. http://dx.doi.org/10.1038/nrn1647.
55. Zhu JJ, Qin Y, Zhao M, Van Aelst L, Malinow R. Ras and Rap control AMPA receptor trafficking during synaptic plasticity. *Cell*. 2002;110(4):443–455. Available at: http://www.ncbi.nlm.nih.gov/pubmed/12202034, Accessed December 2, 2014.
56. Thomas GM, Huganir RL. MAPK cascade signalling and synaptic plasticity. *Nat Rev Neurosci*. 2004;5(3):173–183. http://dx.doi.org/10.1038/nrn1346.
57. Harvey CD, Yasuda R, Zhong H, Svoboda K. The spread of Ras activity triggered by activation of a single dendritic spine. *Science*. 2008;321(5885):136–140. http://dx.doi.org/10.1126/science.1159675.
58. Lynch G, Larson J, Kelso S, Barrionuevo G, Schottler F. Intracellular injections of EGTA block induction of hippocampal long-term potentiation. *Nature*.

1983;305(5936):719–721. Available at: http://www.ncbi.nlm.nih.gov/pubmed/6415483, Accessed March 26, 2013.

59. Lisman J, Yasuda R, Raghavachari S. Mechanisms of CaMKII action in long-term potentiation. *Nat Rev Neurosci.* 2012;13(3):169–182. http://dx.doi.org/10.1038/nrn3192.
60. Patterson MA, Szatmari EM, Yasuda R. AMPA receptors are exocytosed in stimulated spines and adjacent dendrites in a Ras-ERK-dependent manner during long-term potentiation. *Proc Natl Acad Sci USA.* 2010;107(36):15951–15956. http://dx.doi.org/10.1073/pnas.0913875107.
61. Lorente de Nó R. Studies on the structure of the cerebral cortex. II. Continuation of the study of the ammonic system. *J Psychol Neurol.* 1934;46:113–177.
62. Jones MW, McHugh TJ. Updating hippocampal representations: CA2 joins the circuit. *Trends Neurosci.* 2011;34(10):526–535. http://dx.doi.org/10.1016/j.tins.2011.07.007.
63. Caruana DA, Alexander GM, Dudek SM. New insights into the regulation of synaptic plasticity from an unexpected place: hippocampal area CA2. *Learn Mem.* 2012;19(9):391–400. http://dx.doi.org/10.1101/lm.025304.111.
64. Zhao M, Choi Y-S, Obrietan K, Dudek SM. Synaptic plasticity (and the lack thereof) in hippocampal CA2 neurons. *J Neurosci.* 2007;27(44):12025–12032. http://dx.doi.org/10.1523/JNEUROSCI.4094-07.2007.
65. Pizzorusso T, Medini P, Berardi N, Chierzi S, Fawcett JW, Maffei L. Reactivation of ocular dominance plasticity in the adult visual cortex. *Science.* 2002;298(5596):1248–1251. http://dx.doi.org/10.1126/science.1072699.
66. Brückner G, Grosche J, Hartlage-Rübsamen M, Schmidt S, Schachner M. Region and lamina-specific distribution of extracellular matrix proteoglycans, hyaluronan and tenascin-R in the mouse hippocampal formation. *J Chem Neuroanat.* 2003;26(1):37–50. Available at: http://www.ncbi.nlm.nih.gov/pubmed/12954529, Accessed December 2, 2014.
67. Yamamoto M, Marshall P, Hemmendinger LM, Boyer AB, Caviness VS. Distribution of glucuronic acid-and-sulfate-containing glycoproteins in the central nervous system of the adult mouse. *Neurosci Res.* 1988;5(4):273–298. Available at: http://www.ncbi.nlm.nih.gov/pubmed/2453818, Accessed December 2, 2014.
68. Lein ES, Callaway EM, Albright TD, Gage FH. Redefining the boundaries of the hippocampal CA2 subfield in the mouse using gene expression and 3-dimensional reconstruction. *J Comp Neurol.* 2005;485(1):1–10. http://dx.doi.org/10.1002/cne.20426.
69. Lein ES, Hawrylycz MJ, Ao N, et al. Genome-wide atlas of gene expression in the adult mouse brain. *Nature.* 2007;445(7124):168–176. http://dx.doi.org/10.1038/nature05453.
70. Willard FS, Willard MD, Kimple AJ, et al. Regulator of G-protein signaling 14 (RGS14) is a selective H-Ras effector. *PLoS One.* 2009;4(3):e4884. http://dx.doi.org/10.1371/journal.pone.0004884.
71. Seress L, Gulyás AI, Ferrer I, Tunon T, Soriano E, Freund TF. Distribution, morphological features, and synaptic connections of parvalbumin- and calbindin D28k-immunoreactive neurons in the human hippocampal formation. *J Comp Neurol.* 1993;337(2):208–230. http://dx.doi.org/10.1002/cne.903370204.
72. Leranth C, Ribak CE. Calcium-binding proteins are concentrated in the CA2 field of the monkey hippocampus: a possible key to this region's resistance to epileptic damage. *Exp Brain Res.* 1991;85(1):129–136. Available at: http://www.ncbi.nlm.nih.gov/pubmed/1884753, Accessed December 2, 2014.
73. Simons SB, Escobedo Y, Yasuda R, Dudek SM. Regional differences in hippocampal calcium handling provide a cellular mechanism for limiting plasticity. *Proc Natl Acad Sci USA.* 2009;106(33):14080–14084. Available at: http://www.pubmedcentral.nih.gov/articlerender.fcgi?artid=2729023&tool=pmcentrez&rendertype=abstract.

74. Ochiishi T, Saitoh Y, Yukawa A, et al. High level of adenosine A1 receptor-like immunoreactivity in the CA2/CA3a region of the adult rat hippocampus. *Neuroscience*. 1999;93(3):955–967. Available at: http://www.ncbi.nlm.nih.gov/pubmed/10473260, Accessed December 2, 2014.
75. Simons SB, Caruana DA, Zhao M, Dudek SM. Caffeine-induced synaptic potentiation in hippocampal CA2 neurons. *Nat Neurosci*. 2012;15(1):23–25. http://dx.doi.org/10.1038/nn.2962.
76. Arai A, Lynch G. Factors regulating the magnitude of long-term potentiation induced by theta pattern stimulation. *Brain Res*. 1992;598(1–2):173–184. Available at: http://www.ncbi.nlm.nih.gov/pubmed/1486479, Accessed December 18, 2014.
77. Dunwiddie TV, Masino SA. The role and regulation of adenosine in the central nervous system. *Annu Rev Neurosci*. 2001;24:31–55. http://dx.doi.org/10.1146/annurev.neuro.24.1.31.
78. Young WS, Li J, Wersinger SR, Palkovits M. The vasopressin 1b receptor is prominent in the hippocampal area CA2 where it is unaffected by restraint stress or adrenalectomy. *Neuroscience*. 2006;143(4):1031–1039. http://dx.doi.org/10.1016/j.neuroscience.2006.08.040.
79. Pagani JH, Lee H-J, Young WS. Postweaning, forebrain-specific perturbation of the oxytocin system impairs fear conditioning. *Genes Brain Behav*. 2011;10(7):710–719. http://dx.doi.org/10.1111/j.1601-183X.2011.00709.x.
80. Pagani JH, Zhao M, Cui Z, et al. Role of the vasopressin 1b receptor in rodent aggressive behavior and synaptic plasticity in hippocampal area CA2. *Mol Psychiatry*. 2015;20(4):490–499. http://dx.doi.org/10.1038/mp.2014.47.
81. Meyer-Lindenberg A, Domes G, Kirsch P, Heinrichs M. Oxytocin and vasopressin in the human brain: social neuropeptides for translational medicine. *Nat Rev Neurosci*. 2011;12(9):524–538. http://dx.doi.org/10.1038/nrn3044.
82. Kohara K, Pignatelli M, Rivest AJ, et al. Cell type-specific genetic and optogenetic tools reveal hippocampal CA2 circuits. *Nat Neurosci*. 2014;17(2):269–279. http://dx.doi.org/10.1038/nn.3614.
83. Maglóczky Z, Acsády L, Freund TF. Principal cells are the postsynaptic targets of supramammillary afferents in the hippocampus of the rat. *Hippocampus*. 1994;4(3):322–334. http://dx.doi.org/10.1002/hipo.450040316.
84. Chevaleyre V, Siegelbaum SA. Strong CA2 pyramidal neuron synapses define a powerful disynaptic cortico-hippocampal loop. *Neuron*. 2010;66(4):560–572. http://dx.doi.org/10.1016/j.neuron.2010.04.013.
85. Cui Z, Gerfen CR, Young WS. Hypothalamic and other connections with dorsal CA2 area of the mouse hippocampus. *J Comp Neurol*. 2013;521(8):1844–1866. http://dx.doi.org/10.1002/cne.23263.
86. Haglund L, Swanson LW, Köhler C. The projection of the supramammillary nucleus to the hippocampal formation: an immunohistochemical and anterograde transport study with the lectin PHA-L in the rat. *J Comp Neurol*. 1984;229(2):171–185. http://dx.doi.org/10.1002/cne.902290204.
87. Vertes RP. PHA-L analysis of projections from the supramammillary nucleus in the rat. *J Comp Neurol*. 1992;326(4):595–622. http://dx.doi.org/10.1002/cne.903260408.
88. Kiss J, Csáki A, Bokor H, Shanabrough M, Leranth C. The supramammillo-hippocampal and supramammillo-septal glutamatergic/aspartatergic projections in the rat: a combined [3H]D-aspartate autoradiographic and immunohistochemical study. *Neuroscience*. 2000;97(4):657–669. Available at: http://www.ncbi.nlm.nih.gov/pubmed/10842010, Accessed December 2, 2014.
89. Rowland DC, Weible AP, Wickersham IR, et al. Transgenically targeted rabies virus demonstrates a major monosynaptic projection from hippocampal area CA2 to medial

entorhinal layer II neurons. *J Neurosci.* 2013;33(37):14889–14898. http://dx.doi.org/10.1523/JNEUROSCI.1046-13.2013.
90. Llorens-Martín M, Jurado-Arjona J, Avila J, Hernández F. Novel connection between newborn granule neurons and the hippocampal CA2 field. *Exp Neurol.* 2015;263:285–292. http://dx.doi.org/10.1016/j.expneurol.2014.10.021.
91. Burgalossi A, Herfst L, Von Heimendahl M, et al. Microcircuits of functionally identified neurons in the rat medial entorhinal cortex. *Neuron.* 2011;70(4):773–786. http://dx.doi.org/10.1016/j.neuron.2011.04.003.
92. Couey JJ, Witoelar A, Zhang S-J, et al. Recurrent inhibitory circuitry as a mechanism for grid formation. *Nat Neurosci.* 2013;16(3):318–324. http://dx.doi.org/10.1038/nn.3310.
93. Schmidt-Hieber C, Häusser M. Cellular mechanisms of spatial navigation in the medial entorhinal cortex. *Nat Neurosci.* 2013;16(3):325–331. http://dx.doi.org/10.1038/nn.3340.
94. DeVito LM, Konigsberg R, Lykken C, Sauvage M, Young WS, Eichenbaum H. Vasopressin 1b receptor knock-out impairs memory for temporal order. *J Neurosci.* 2009;29(9):2676–2683. http://dx.doi.org/10.1523/JNEUROSCI.5488-08.2009.
95. Serradeil-Le Gal C, Wagnon J, Tonnerre B, et al. An overview of SSR149415, a selective nonpeptide vasopressin V(1b) receptor antagonist for the treatment of stress-related disorders. *CNS Drug Rev.* 2005;11(1):53–68. Available at: http://www.ncbi.nlm.nih.gov/pubmed/15867952, Accessed December 19, 2014.
96. Hitti FL, Siegelbaum SA. The hippocampal CA2 region is essential for social memory. *Nature.* 2014;508(7494):88–92. http://dx.doi.org/10.1038/nature13028.
97. Allen Brain Institute. *Allen Human Brain Atlas.* Available at: http://human.brain-map.org/.
98. *NIH Blueprint Non-Human Primate (NHP) Atlas.* Available at: http://www.blueprintnhpatlas.org/.
99. Kirino T. Delayed neuronal death in the gerbil hippocampus following ischemia. *Brain Res.* 1982;239(1):57–69. Available at: http://www.ncbi.nlm.nih.gov/pubmed/7093691, Accessed March 26, 2013.
100. Sadowski M, Wisniewski HM, Jakubowska-Sadowska K, Tarnawski M, Lazarewicz JW, Mossakowski MJ. Pattern of neuronal loss in the rat hippocampus following experimental cardiac arrest-induced ischemia. *J Neurol Sci.* 1999;168(1):13–20. Available at: http://www.ncbi.nlm.nih.gov/pubmed/10500268, Accessed March 26, 2013.
101. Benes FM, Kwok EW, Vincent SL, Todtenkopf MS. A reduction of nonpyramidal cells in sector CA2 of schizophrenics and manic depressives. *Biol Psychiatry.* 1998;44(2):88–97. Available at: http://www.ncbi.nlm.nih.gov/pubmed/9646890, Accessed March 26, 2013.
102. Dam AM. Epilepsy and neuron loss in the hippocampus. *Epilepsia.* 1980;21(6):617–629. Available at: http://www.ncbi.nlm.nih.gov/pubmed/6777154, Accessed December 2, 2014.
103. Ishizawa T, Ko L, Cookson N, Davias P, Espinoza M, Dickson DW. Selective neurofibrillary degeneration of the hippocampal CA2 sector is associated with four-repeat tauopathies. *J Neuropathol Exp Neurol.* 2002;61(12):1040–1047. Available at: http://www.ncbi.nlm.nih.gov/pubmed/12484566, Accessed November 26, 2014.
104. Benes FM, Sorensen I, Bird ED. Reduced neuronal size in posterior hippocampus of schizophrenic patients. *Schizophr Bull.* 1991;17(4):597–608. Available at: http://www.ncbi.nlm.nih.gov/pubmed/1805353, Accessed December 18, 2014.
105. Gao X-M, Sakai K, Roberts RC, Conley RR, Dean B, Tamminga CA. Ionotropic glutamate receptors and expression of N-methyl-D-aspartate receptor subunits in

subregions of human hippocampus: effects of schizophrenia. *Am J Psychiatry*. 2000;157(7):1141–1149. http://dx.doi.org/10.1176/appi.ajp.157.7.1141.
106. Maxwell WL, Dhillon K, Harper L, et al. There is differential loss of pyramidal cells from the human hippocampus with survival after blunt head injury. *J Neuropathol Exp Neurol*. 2003;62(3):272–279. Available at: http://www.ncbi.nlm.nih.gov/pubmed/12638731, Accessed December 2, 2014.
107. Sloviter RS. "Epileptic" brain damage in rats induced by sustained electrical stimulation of the perforant path. I. Acute electrophysiological and light microscopic studies. *Brain Res Bull*. 1983;10(5):675–697. Available at: http://www.ncbi.nlm.nih.gov/pubmed/6871737, Accessed December 2, 2014.
108. Vigers AJ, Baquet ZC, Jones KR. Expression of neurotrophin-3 in the mouse forebrain: insights from a targeted LacZ reporter. *J Comp Neurol*. 2000;416(3):398–415. Available at: http://www.ncbi.nlm.nih.gov/pubmed/10602097, Accessed December 2, 2014.
109. Cheng B, Mattson MP. NT-3 and BDNF protect CNS neurons against metabolic/excitotoxic insults. *Brain Res*. 1994;640(1–2):56–67. Available at: http://www.ncbi.nlm.nih.gov/pubmed/7911729, Accessed December 2, 2014.
110. Young D, Dragunow M. Neuronal injury following electrically induced status epilepticus with and without adenosine receptor antagonism. *Exp Neurol*. 1995;133(2):125–137. http://dx.doi.org/10.1006/exnr.1995.1015.
111. Williamson A, Spencer DD. Electrophysiological characterization of CA2 pyramidal cells from epileptic humans. *Hippocampus*. 1994;4(2):226–237. http://dx.doi.org/10.1002/hipo.450040213.
112. Sloviter RS. Permanently altered hippocampal structure, excitability, and inhibition after experimental status epilepticus in the rat: the "dormant basket cell" hypothesis and its possible relevance to temporal lobe epilepsy. *Hippocampus*. 1991;1(1):41–66. http://dx.doi.org/10.1002/hipo.450010106.
113. Sloviter RS. Calcium-binding protein (calbindin-D28k) and parvalbumin immunocytochemistry: localization in the rat hippocampus with specific reference to the selective vulnerability of hippocampal neurons to seizure activity. *J Comp Neurol*. 1989;280(2):183–196. http://dx.doi.org/10.1002/cne.902800203.
114. Winn HR, Welsh JE, Rubio R, Berne RM. Changes in brain adenosine during bicuculline-induced seizures in rats. Effects of hypoxia and altered systemic blood pressure. *Circ Res*. 1980;47(4):568–577. Available at: http://www.ncbi.nlm.nih.gov/pubmed/6773698, Accessed November 20, 2014.
115. Laeremans A, Nys J, Luyten W, D'Hooge R, Paulussen M, Arckens L. AMIGO2 mRNA expression in hippocampal CA2 and CA3a. *Brain Struct Funct*. 2013;218(1):123–130. http://dx.doi.org/10.1007/s00429-012-0387-4.
116. Ono T, Sekino-Suzuki N, Kikkawa Y, Yonekawa H, Kawashima S. Alivin 1, a novel neuronal activity-dependent gene, inhibits apoptosis and promotes survival of cerebellar granule neurons. *J Neurosci*. 2003;23(13):5887–5896. Available at: http://www.ncbi.nlm.nih.gov/pubmed/12843293, Accessed December 2, 2014.
117. Caldwell HK, Wersinger SR, Young WS. The role of the vasopressin 1b receptor in aggression and other social behaviours. *Prog Brain Res*. 2008;170:65–72. http://dx.doi.org/10.1016/S0079-6123(08)00406-8.
118. Parker CC, Sokoloff G, Cheng R, Palmer AA. Genome-wide association for fear conditioning in an advanced intercross mouse line. *Behav Genet*. 2012;42(3):437–448. http://dx.doi.org/10.1007/s10519-011-9524-8.
119. Fiala JC, Feinberg M, Popov V, Harris KM. Synaptogenesis via dendritic filopodia in developing hippocampal area CA1. *J Neurosci*. 1998;18(21):8900–8911. Available at: http://www.ncbi.nlm.nih.gov/pubmed/9786995, Accessed December 2, 2014.

120. San Antonio A, Liban K, Ikrar T, Tsyganovskiy E, Xu X. Distinct physiological and developmental properties of hippocampal CA2 subfield revealed by using anti-Purkinje cell protein 4 (PCP4) immunostaining. *J Comp Neurol.* 2014;522(6):1333–1354. http://dx.doi.org/10.1002/cne.23486.
121. Dityatev A, Schachner M, Sonderegger P. The dual role of the extracellular matrix in synaptic plasticity and homeostasis. *Nat Rev Neurosci.* 2010;11(11):735–746. http://dx.doi.org/10.1038/nrn2898.
122. Wilson DA, Sullivan RM. Cortical processing of odor objects. *Neuron.* 2011;72(4):506–519. http://dx.doi.org/10.1016/j.neuron.2011.10.027.
123. López-Aranda MF, López-Téllez JF, Navarro-Lobato I, Masmudi-Martín M, Gutiérrez A, Khan ZU. Role of layer 6 of V2 visual cortex in object-recognition memory. *Science.* 2009;325(5936):87–89. http://dx.doi.org/10.1126/science.1170869.
124. Landrum MJ, Lee JM, Riley GR, et al. ClinVar: public archive of relationships among sequence variation and human phenotype. *Nucleic Acids Res.* 2014;42(Database issue): D980–D985. http://dx.doi.org/10.1093/nar/gkt1113.

INDEX

Note: Page numbers followed by "*f*" indicate figures and "*t*" indicate tables.

H

I

K

L

M

N

O

P

R

S

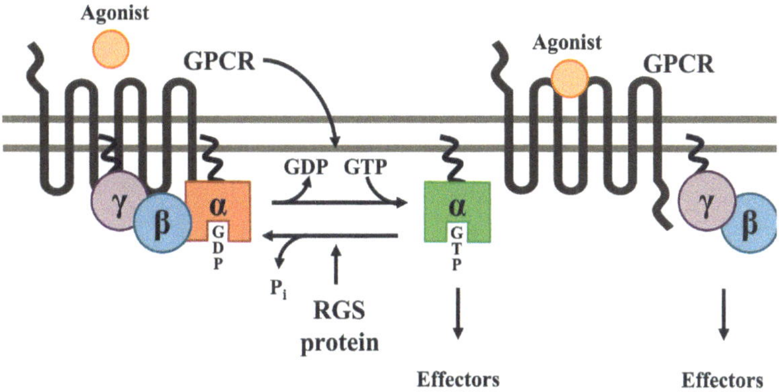

Adele Stewart and Rory A. Fisher, Figure 1 Canonical regulation of GPCR signaling by RGS proteins. Agonist binding to GPCRs induces a conformation change that facilitates the exchange of GDP for GTP on the α subunit of the heterotrimeric complex. Both GTP-bound Gα in the active form and the released Gβγ dimer can then go on to stimulate a number of downstream effectors. RGS proteins are GAPs for Gα, which function to terminate signaling through GPCRs by accelerating the intrinsic GTPase activity of Gα and promoting reassociation of the heterotrimeric complex with the receptor at the cell membrane.

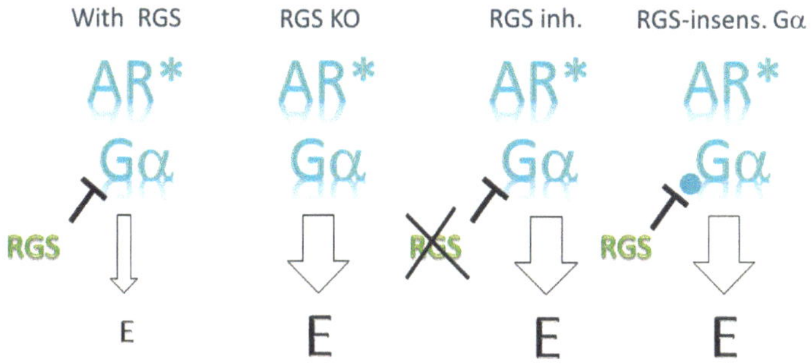

Richard R. Neubig, Figure 1 Comparison of distinct approaches to disrupt RGS protein functions. Agonist-activated receptors AR* induce G protein activation to produce functional effects (E). In the presence of an active RGS protein, the effect is smaller (smaller E) than in its absence (second model). RGS proteins can be knocked out themselves disrupting the action of only that RGS protein and revealing its specific actions. Redundancy with multiple RGS proteins present in the cell (cardiac cells express >10 different RGS protein mRNAs) often lead to modest effects. In contrast, the G184S mutant of the Gα subunit (last model) prevents the actions of all RGS proteins. As noted in Table 1, this provides a more robust idea of the role of RGS proteins overall but the specificity of the effects is that of the Gα subunit and not of any individual RGS protein.

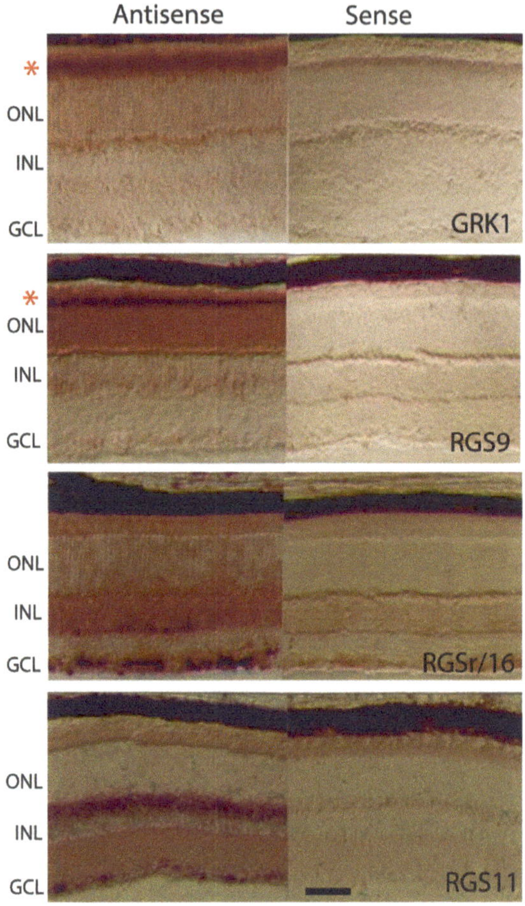

Ching-Kang Jason Chen, Figure 1 *In situ* hybridization showing localization of GRK1, RGS9, RGSr/16, and RGS11 messages in mouse retinal cross sections. The three nuclear layers: outer nuclear layer (ONL), inner nuclear layer (INL), and ganglion cell layer (GCL) are marked at left. GRK1 message is abundantly present at the inner segment (IS) layer of the retina (red asterisk) and is used as a marker for any potential photoreceptor-specific genes. Sense probes control for degree of background staining. For RGS genes tested for candidacy as a transducin GAP, only RGS9 message appears in the IS and ONL. RGSr/16 is present throughout INL and GCL but notably absent in IS or ONL. RGS11 is localized to some GCL cells and the upper-tier cells in INL. Scale bar equals 50 μm.

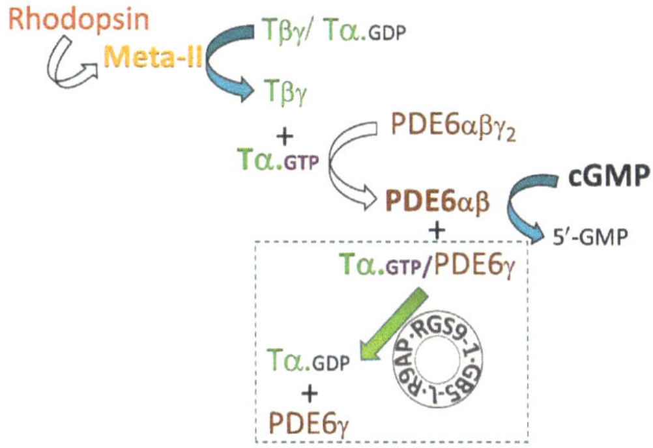

Ching-Kang Jason Chen, Figure 2 Vertebrate rod phototransduction cascade. Photon absorption by rhodopsin leads to the formation of metarhodopsin II (Meta-II), which catalyzes guanine nucleotide exchange on transducin α subunit (Tα) and results in its dissociation from the βγ dimer (Tβγ). The GTP-bound Tα (Tα.GTP) interacts with PDE6γ to relieve its inhibition on PDE6αβ, freeing this near-perfect phosphodiesterase to hydrolyze cGMP into 5′-GMP. Two gain steps marked by filled ribbons endow rods with the sensitivity to detect single photons. This chapter deals with one reaction that must occur during recovery of rod phototransduction as highlighted in the dashed box. The reaction is the hydrolysis of bound GTP by Tα to GDP, which is assisted *in vivo* by a GTPase-accelerating protein (GAP), now known to be a protein complex consists of RGS9-1, Gβ5-L, and R9AP. Without the GAP complex, Tα turns over bound GTP in seconds, a pace too slow to sustain normal vision.

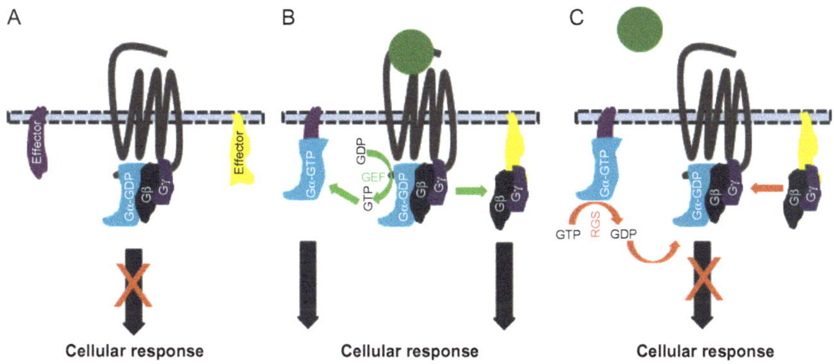

Joel Jules *et al.*, Figure 1 Schematic of GPCR–G protein–RGS activation and inactivation cycle. (A) GPCR inactivation in the absence of ligands. In absence of ligands, Gα is linked to GDP (Gα-GDP) and forms a heterotrimeric complex with the Gγ and Gβ subunits, preventing their liberation to activate effectors to induce cellular response. (B) Ligand-induced GPCR activation. In the presence of ligands, GPCR undergoes a conformational change leading to the activation of Gα through the exchange of GDP for GTP by the receptor GEF activity. Gα-GTP and Gβγ subunits dissociate to modulate the activity of specific effectors to mediate cellular responses. (C) GPCR signaling inactivation by RGS proteins. RGS proteins enhance the GTPase activity of Gα to hydrolyze GTP to GDP leading to the inactivation of the Gα protein and its subsequent reunification with Gβγ subunits to reform the inactive G protein heterotrimeric complex. This prevents the G proteins from continuing activating effectors for cellular response and thus terminates the cellular response. GDP, guanosine diphosphate; GTP, guanosine triphosphate; GEF, guanine nucleotide exchange factor.

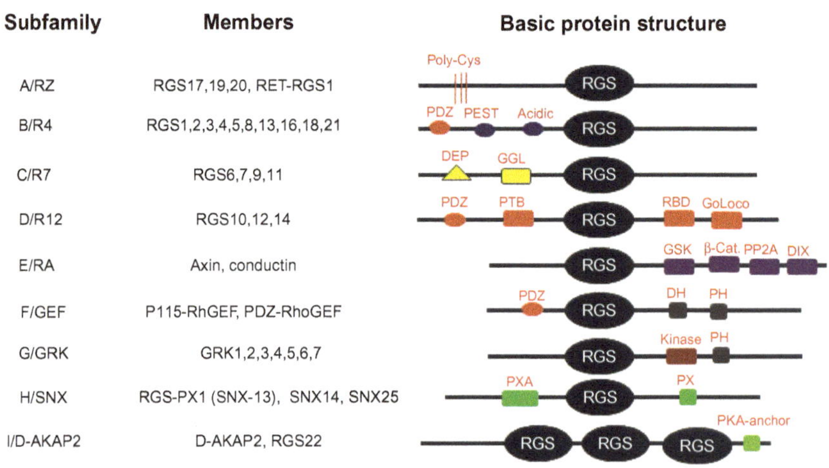

Joel Jules *et al.*, Figure 2 Subclassification of RGS family members. All members of the RGS family possess a common RGS domain but are subdivided into different subfamilies based on protein structures. β-Cat, β-catenin-binding; D-AKAP, dual-specificity A-kinase anchoring protein; DEP, dishevelled/EGL-10/pleckstrin; DH, double homology; DIX, dishevelled homology domain; GEF, guanine nucleotide exchange factor; GGL, Gγ-like; Goloco, $G\alpha_{i/o}$-Loco; GRK, G protein-coupled receptor kinase; GSK, glycogen synthase kinase 3β-binding; PDZ, PSD95/dlg/Z0-1/2; PEST, proline, glutamine, serine, threonine-rich; PH, pleckstrin homology; PP2A, protein phosphatase 2A; PTB, phosphotyrosine-binding; PX, phosphatidylinositol-binding; PXA, PX-associated; RBD, Ras-binding domain; SNX, sorting nexin.

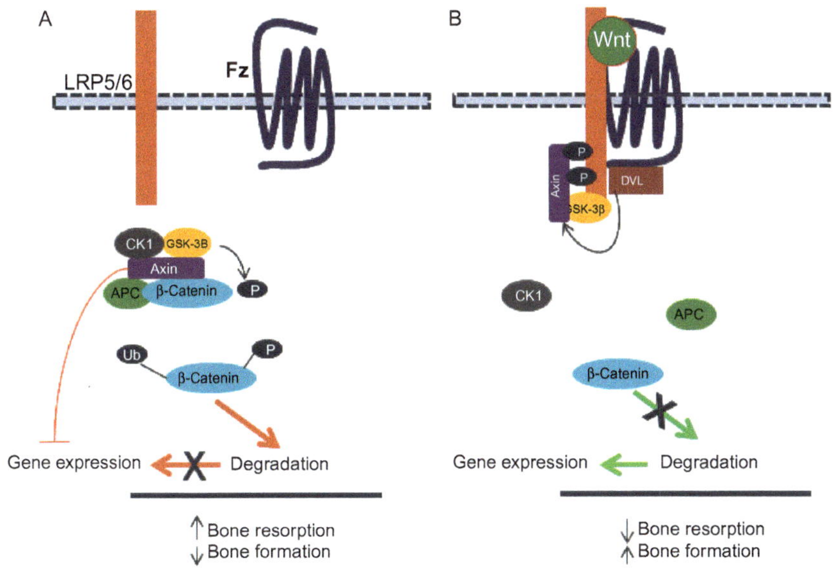

Joel Jules *et al.*, Figure 3 Role of axin in Wnt signaling in bone. (A) In the absence of Wnt ligands, β-catenin is constructively targeted for degradation via a multiprotein destruction complex. β-Catenin is phosphorylated by GSK-3β and then targeted for ubiquitination (Ub) and proteosomal degradation. Axin plays a key role in this process of β-catenin degradation and thus functions as an inhibitor of the canonical Wnt-β pathway. (B) Binding of canonical Wnt ligands to their co-receptors Fz and LRP5/6 prevent β-catenin phosphorylation by GSK-3β, which in turn prevent its degradation and allow it to move to the nucleus to mediate gene expression.

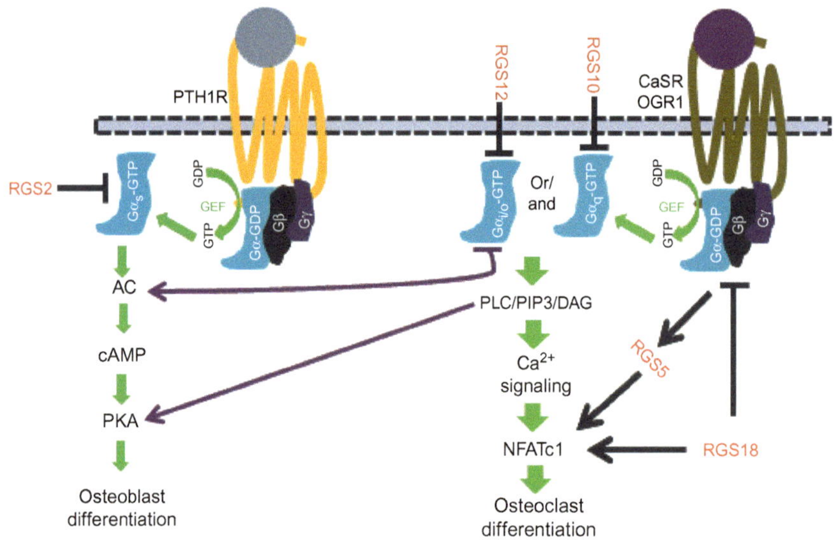

Joel Jules *et al.*, Figure 4 Role of RGS proteins in selected GPCR signaling in bone. In osteoblasts, PTH1R activation upon binding of PTH or PTHrP triggers the activation of the $G\alpha_s$–AC–cAMP signaling pathway which ultimately promotes osteoblast differentiation. Activation of PTH1R can induce the expression of many RGS proteins, including RGS2 and RGS5. RGS2 can target the cAMP signaling to regulate PTH1R signaling, and RGS5 is likely to regulate osteoblast differentiation by controlling PTH levels. In osteoclasts, activation of GPCRs, such as CasR and OGR1, can activate $G\alpha_q$ or $G\alpha_{i/o}$ to activate the PLC–Ca^{2+} signaling pathway for NFATc1 activation or $G\alpha_s$ to activate the Ac–cAMP pathway to promote osteoblast differentiation and function. RGS10, RGS12, and RGS18 can regulate osteoclast differentiation.

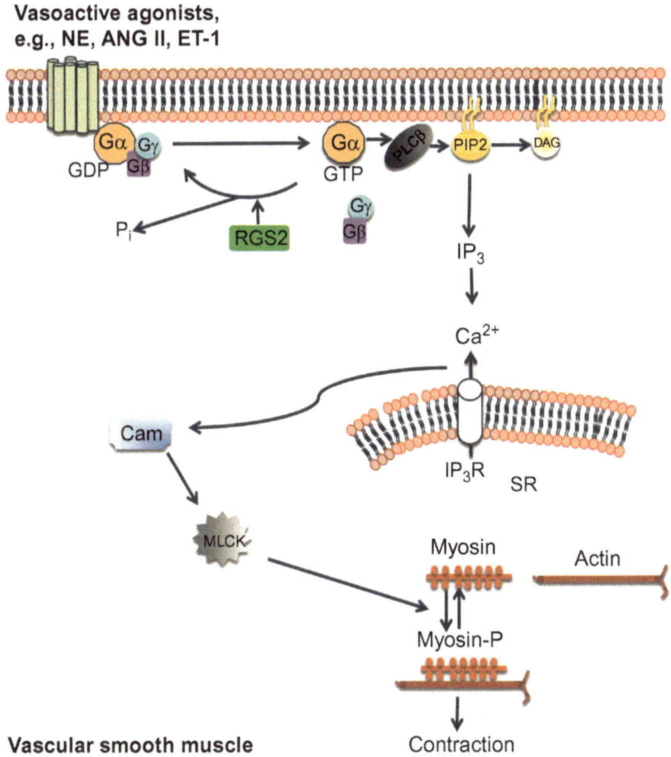

Patrick Osei-Owusu and Kendall J. Blumer, Figure 1 The role of RGS2 in vascular smooth muscle contraction. RGS2 tightly regulates the duration and amplitude of G protein signaling that mediates smooth muscle contraction triggered by vasoactive ligands that activate G_q-coupled GPCRs. Upon activation, G_q signaling evokes a rise in intracellular Ca^{2+} concentration resulting in myosin phosphorylation by myosin light chain kinase (MLCK) and actomyosin crossbridge formation. Decreased RGS2 expression or function can lead to increased signaling duration and intensity that in turn cause augmented smooth muscle contraction. Cam, calmodulin; DAG, diacylglycerol; PIP_2, phosphatidylinositol 4,5-bisphosphate; PLCβ, phospholipase C-beta; IP_3, inositol 1,3,5-trisphosphate; SR, sarcoplasmic reticulum; IP_3R, IP_3 receptor; NE, norepinephrine; ET-1, endothelin-1; ANG II, angiotensin II.

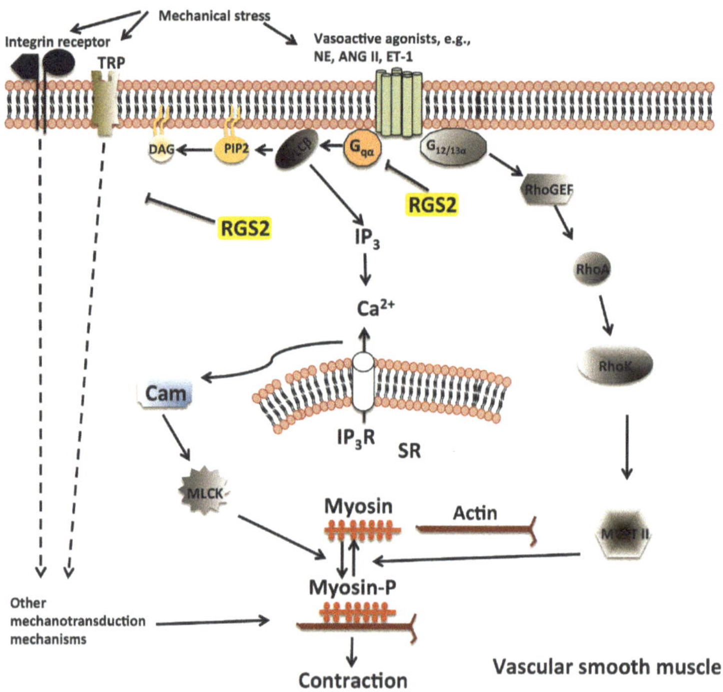

Patrick Osei-Owusu and Kendall J. Blumer, Figure 2 A model of contractile signaling pathways potentially regulated by RGS2 in smooth muscle cells in the resistance vasculature. In addition to regulating G_q-dependent smooth muscle contraction, RGS2 could potentially regulate stretch-induced contraction mediated by putative receptors and ion channels that can be activated by mechanical stimuli. TRP, transient receptor potential; NE, norepinephrine; ET-1, endothelin-1; ANG II, angiotensin II; myosin-P, phosphorylated myosin; RhoK, RhoA-associated kinase; MYPT II, myosin phosphatase regulatory subunit II.

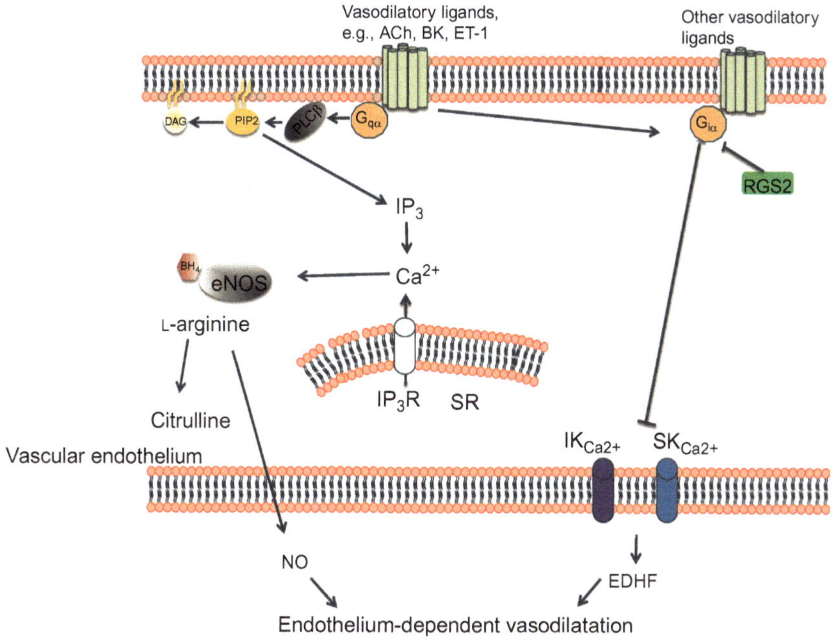

Patrick Osei-Owusu and Kendall J. Blumer, Figure 3 Regulation of G protein signaling in vascular endothelium. Vasodilatory ligands such as ACh and BK activate G_q-coupled receptors in the vascular endothelium of resistance arteries causing a rise in intracellular Ca^{2+}, required to stimulate endothelial nitric oxide synthase (eNOS) for producing nitric oxide (NO). A rise in endothelial Ca^{2+} is also necessary for activation of potassium channels SK_{Ca} and IK_{Ca} that mediate the production of endothelium-derived hyperpolarizing factor (EDHF). The activity of SK_{Ca} and IK_{Ca} are negatively regulated by Gα subunits of the G_i-class in this vascular compartment. By acting as GAP, RGS2 squelches the activity of G_i-class G proteins to promote EDHF production. Thus, RGS2 deficiency in the endothelium is proposed to result in impairment of endothelium-dependent vasodilatation due to increased G_i signaling, inhibiting EDHF production. IP_3, inositol 1,4,5-trisphosphate; PLCβ, phospholipase C-beta; BH_4, tetrahydrobiopterin; eNOS, endothelium nitric oxide synthase.

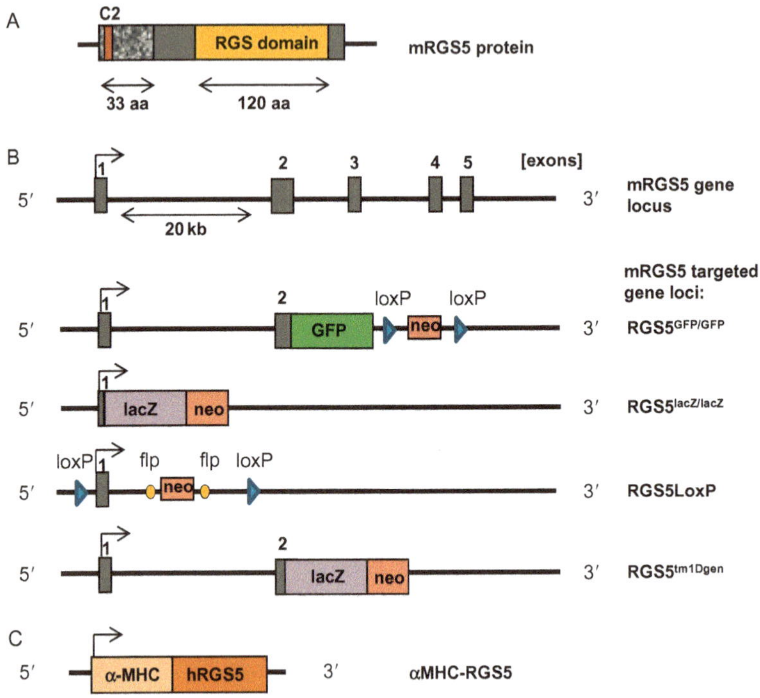

Ruth Ganss, Figure 1 Murine RGS5 protein/gene structures and RGS5-targeted gene loci. (A) Murine RGS5 protein consists of a 120 aa conserved RGS domain which confers GTPase activity and a 33 aa N-terminus which contains a cysteine residue at position 2 (C2) and a potential amphipathic helix domain. (B) The murine RGS5 gene comprises 5 exons. Gene deletion analysis demonstrates that intron 1 (approximately 20 kb) contains important regulatory elements[19] and a functional PPARγ and β/δ binding site.[20] RGS5 knockout mouse strains were generated using the following strategies: Nisancioglu *et al.* inserted a green fluorescent protein (GFP) and a neomycin resistance gene flanked with LoxP sites in frame with exon 2. The selection cassette (neo) was removed using Cre/LoxP technology ($RGS5^{GFP/GFP}$).[21] Cho *et al.* generated knockout mice by cloning a lacZ/neomycin gene cassette (LacZ neo) in frame with exon 1. This strategy produced RGS5-deficient mice, but the lacZ gene was inactive and the neo cassette was not removed ($RGS5^{LacZ/LacZ}$).[19] Hamzah *et al.* flanked exon 1 with LoxP sites. The neo cassette was removed by flp recombination (RGS5LoxP). RGS5-deficient mice were generated by crossing RGS5-LoxP mice with Cre-deleter mice.[22] Another LacZ/neo expressing RGS5 knockout strain was generated by Deltagen Inc. In this line, lacZ/neo was inserted into exon 2 and is expressed from the endogenous RGS5 promoter ($RGS5^{tm1Dgen}$). (C) A cardiac myocyte-specific, RGS5 overexpressing mouse line was generated by cloning the human cardiac myosin heavy chain (αMHC) promoter in front of the human RGS5 cDNA (αMHC-RGS5).[23]

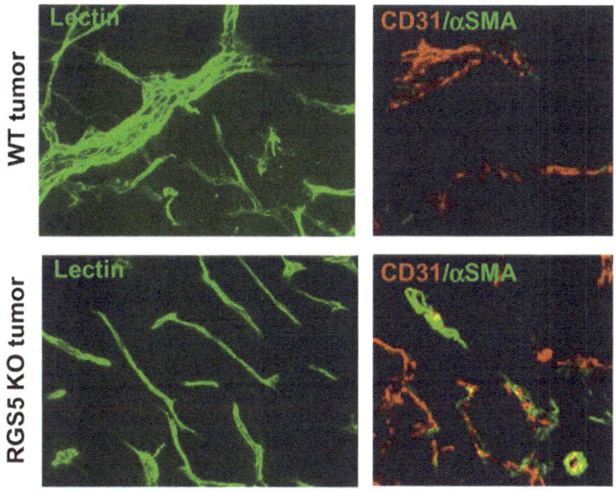

Ruth Ganss, Figure 2 RGS5-deficient tumor vessels are normalized. Confocal images showing tumor vessels in a fibrosarcoma grown in wild-type mice. Lectin-FITC perfused vessels with typical heterogeneous vessel caliber are depicted (upper left panel). In contrast, vessels from tumors grown in RGS5 knockout mice are homogeneous in caliber (lower left panel, magnification 20 ×). Blood vessels (CD31 staining in red) in wild-type tumors are immature with few αSMA-positive pericytes (αSMA staining in green, upper right panel). Normalized blood vessels in RGS5 knockout mice show increased coverage by αSMA-positive pericytes which stabilize vessels (lower left panel, magnification 60 ×).[22]

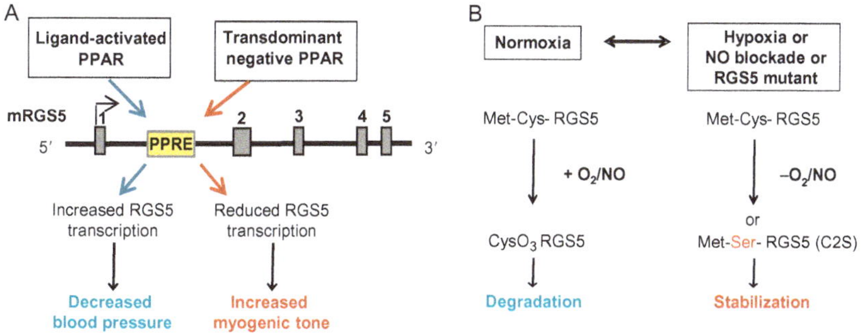

Ruth Ganss, Figure 3 Schematic representation of transcriptional and proteasomal regulation of RGS5. (A) The RGS5 gene contains an active PPAR response element (PPRE) within intron 1.[106] Ligand-activated PPARβ/δ induces RGS5 mRNA in aorta of spontaneously hypertensive rats concomitant with a reduction in blood pressure.[102] Overexpression of transdominant negative PPARγ specifically in blood vessels decreases RGS5 mRNA and increases myogenic tone in mice.[106] (B) Posttranslational control of RGS5 levels. The N-terminal cysteine residue (Cys, C2) in the RGS5 protein is essential for the ubiquitin-proteasome-dependent N-end rule degradation pathway. Specifically, NO-mediated oxidation of N-terminal Cys ($CysO_3$) is required for its arginylation by arginyl transferase (ATE-1). Arginylation then leads to degradation by E3 ubiquitin ligases. Mutation of Cys to Ser at position 2 in RGS5 stabilizes the protein. Met, methionine; Cys, cysteine; Ser, serine; O_2, oxygen; NO, nitric oxide. *Modified from Ref. 98.*

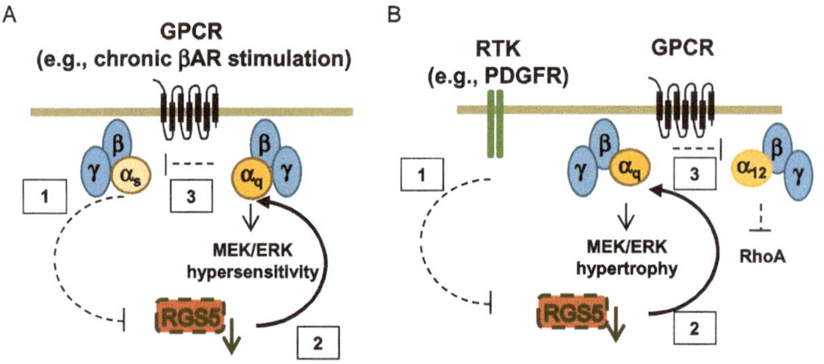

Ruth Ganss, Figure 4 Hypothetical role of RGS5 as modulator between different G protein subtypes and receptor prototypes. (A) Chronic stimulation of Gαs-conjugated β adrenergic receptor in airway smooth muscle cells results in downregulation of RGS5 and increased Gαq-mediated signaling which leads to aSMC hyper-responsiveness.[55] (B) PDGFR activation following vascular injury downregulates RGS5 which in turn enhances GPCR signaling via Gαq. The net outcome of sustained injury is increased vSMC migration/proliferation and hypertrophy.[100] Increase in Gαq signaling may decrease Gα12/13 and RhoA activation.[48]

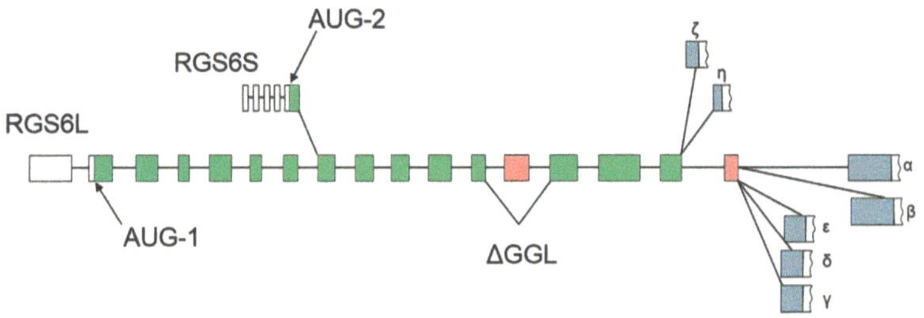

Adele Stewart *et al.*, Figure 1 Schematic depicting the splicing events occurring in the human *RGS6* gene to generate the 36 known RGS6 isoforms. Exons are shown as filled boxes and noncoding sequences as empty boxes. Two primary transcripts with alternate translation start sites (AUG-1 and AUG-2) generate RGS6L or RGS6S forms, the latter lacking a large portion of the N-terminus including the disheveled, EGL-10, pleckstrin homology/DEP helical extension (DEP/DHEX) domains. Retention or skipping of exon 13 generates transcripts with or without a GGL domain (ΔGGL), respectively. Extensive 3′ splicing generates seven different 3′ ends. Use of alternate splice sites in exon 20 generates α and β species while γ, ε, and δ arise from the use of three alternate sites in exon 19. Splicing of exon 16 or 17 to an alternate splice site in exon 18 generates η and ζ, respectively. Additionally, α and β isoforms may contain (1) or lack (2) exon 17. *Image adapted from Ref. 25.*

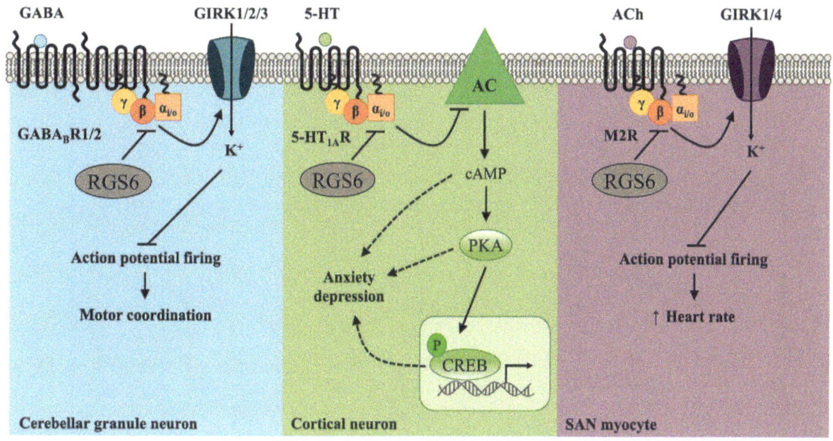

Adele Stewart *et al.*, Figure 3 Schematic outlining the role of RGS6 in GPCR signaling. RGS6 functions to limit the extent of GABA$_B$R-mediated GIRK channel activation in cerebellar granule cells; 5-HT$_{1A}$R-mediated inhibition of AC in cortical neurons and M2R-mediated GIRK current in SAN myocytes. In this way, RGS6 controls motor behavior, baseline anxiety and depression, and heart rate *in vivo*.

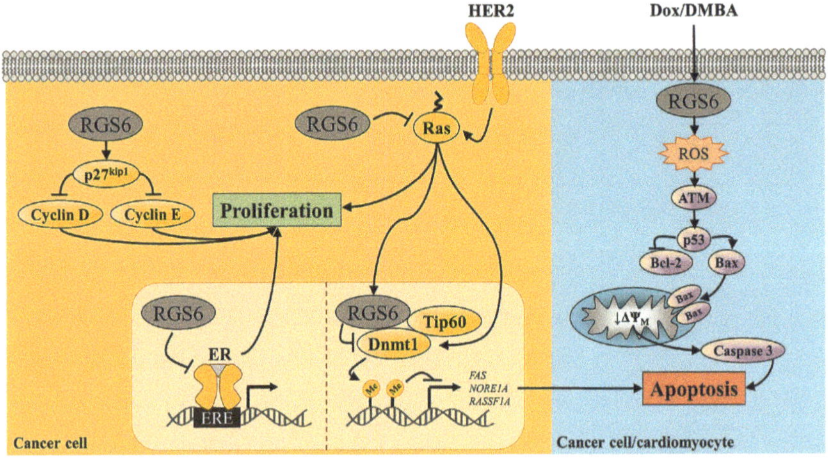

Adele Stewart *et al.*, Figure 4 Schematic outlining the role of RGS6 in G protein-independent signaling. In cancer cells, RGS6 is a multifunctional tumor suppressor capable of simultaneously promoting cell cycle arrest and blocking progrowth signals downstream of HER2 and ER. Further, the ability of RGS6 to promote Tip60-mediated Dnmt1 acetylation and degradation is essential to combating Ras-induced oncogenesis. In addition, RGS6 is a critical upstream modulator of ROS-dependent activation of the ATM/p53/apoptosis pathway in response to doxorubicin in cancer cells and heart.

A

hRGS10L(181aa)

MFNRAVSRLS RKRPPSDIHD SDGSSSSSHQ SLKSTAKWAA SLENLLEDPE
GVKRFREFLK KEFSEENVLF WLACEDFKKM QDKTQMQEKA KEIYMTFLSS
KASSQVNVEG QSRLNEKILE EPHPLMFQKL QDQIFNLMKY DSYSRFLKSD
LFLKHKRTEE EEEDLPDAQT AAKRASRIYN T

hRGS10(173aa)

MQSELCFADI HDSDGSSSSS HQSLKSTAKW AASLENLLED PEGVKRFREF
LKKEFSEENV LFWLACEDFK KMQDKTQMQE KAKEIYMTFL SSKASSQVNV
EGQSRLNEKI LEEPHPLMFQ KLQDQIFNLM KYDSYSRFLK SDLFLKHKRT
EEEEEDLPDA QTAAKRASRI YNT

hRGS10S(167aa)

MEHIHDSDGS SSSSHQSLKS TAKWAASLEN LLEDPEGVKR FREFLKKEFS
EENVLFWLAC EDFKKMQDKT QMQEKAKEIY MTFLSSKASS QVNVEGQSRL
NEKILEEPHP LMFQKLQDQI FNLMKYDSYS RFLKSDLFLK HKRTEEEEED
LPDAQTAAKR ASRIYNT

B

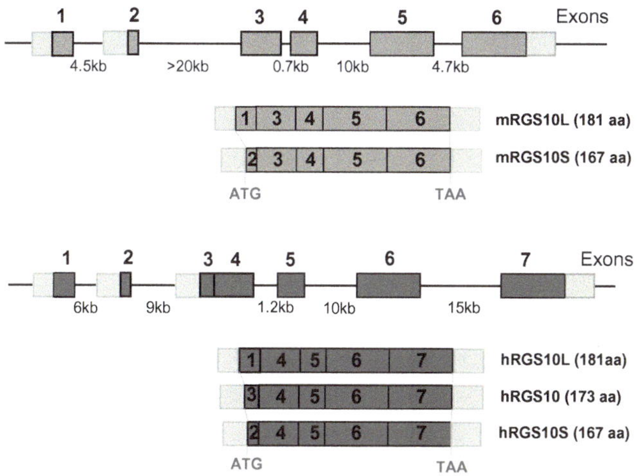

Jae-Kyung Lee and Malú G. Tansey, Figure 1 Sequences of human RGS10 isoforms and structures of the mRNA encoded by *Rgs10* gene. (A) Sequence of human RGS10 isoform polypeptides. The sequence data are available from NCBI Web site under accession numbers NM_001005339 (hRGS10L), AF368902 (hRGS10), and NM_002925 (hRGS10S). The "KRAS motif," cAMP-dependent protein kinase A (PKA) phosphorylation sites are in red (or underlined). The differences in the N-terminus between isoforms as results of splicing are in blue. (B) Structure of the mRNA encoded by the mouse and human *RGS10* genes. The mouse *Rgs10* gene is composed of six exons, while the human *Rgs10* gene is composed of seven exons. The differences between the isoforms occur only in the N-terminal sequences as a result of splicing events in the first two exons in mouse or the first three exons in human.

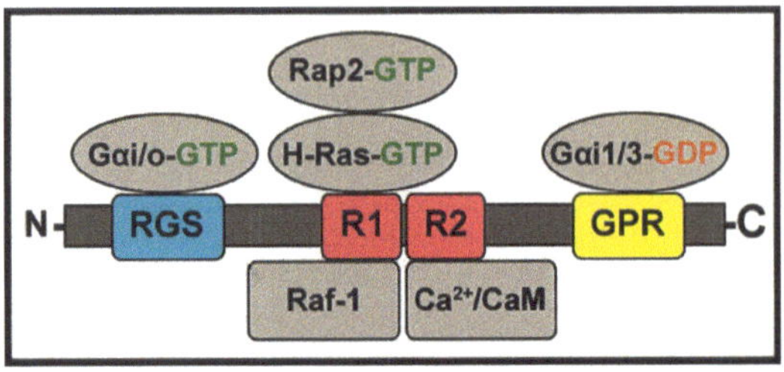

Paul R. Evans *et al.*, Figure 1 RGS14 domain structure and identified binding partners. RGS14 directly binds activated Gαo and Gαi subunits through its N-terminal RGS domain and selectively interacts with inactive Gαi1 and Gαi3 at the GPR motif. RGS14 also contains tandem Ras/Rap-binding domains (R1 and R2) that directly bind activated H-Ras, Rap2, Ca^{2+}/CaM, and Raf kinases.

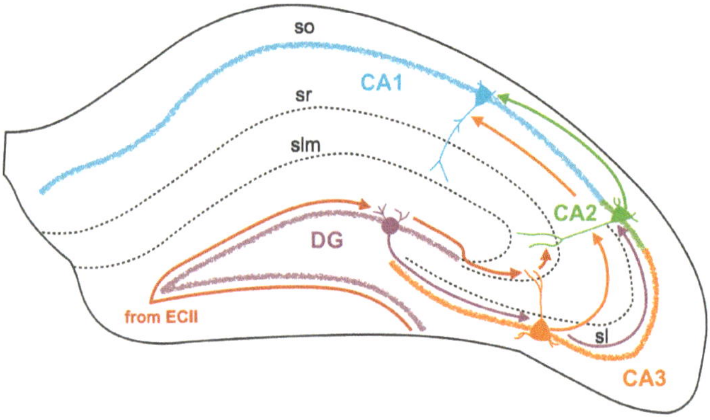

Paul R. Evans *et al.*, Figure 3 Area CA2 at the intersection of multiple hippocampal circuits. CA2 pyramidal neurons receive inputs from different brain regions along the proximal–distal axis of its dendrites (green). Hippocampal inputs from entorhinal cortex layer II (ECII, red) form synapses onto granule cells in the dentate gyrus (DG) and distal dendrites of CA3 and CA2 pyramidal neurons in the stratum lacunosum moleculare (slm). DG granule cells send mossy fiber terminals through the stratum lucidum (sl) that synapse onto the proximal dendrites of CA3 and CA2 pyramidal neurons (purple). CA3 pyramidal neurons form Schaffer collateral synapses (orange) onto CA2, which are usually LTP resistant, as well as CA1 in the intermediate region of the dendrite in the stratum radiatum (sr). CA2 neurons form strong monosynaptic connections onto CA1 pyramidal neurons in the stratum oriens (so).

FSC
www.fsc.org
MIX
Papier | Fördert gute Waldnutzung
FSC® C083411

Kolibri 360 GmbH
Ferdinand-Jühlke-Straße 7
99095 Erfurt, Deutschland
produktsicherheit@kolibri360.de